General College Biology
Laboratory Manual

Second Edition

Jean deSchweinitz
Ping-sha Sheffield
Shawn Snaples
Larry Collins
Tarrant County College

Kendall Hunt
publishing company

CONTENTS

ACKNOWLEDGMENTS

General College Biology Laboratory Manual is a team effort. The four authors worked together to design a lab manual for general college Biology students on the South Campus of Tarrant County College, Ft. Worth, Texas. Jean deSchweinitz wishes to recognize the expertise, hard work, and collaboration of Mr. Collins, Ms. Sheffield, and Ms. Snaples.

The authors wish to recognize Mr. Sam Hill, the technical expert in the Department of Biology, and the Department of Physical Sciences, for his design on the cover of the lab manual (www.imalloverthemap.com).

The authors wish to acknowledge Mr. Robert "Bob" Nabors, Professor Emeritus of Biology, Tarrant County College. Jean deSchweinitz has known and admired Mr. Nabors for almost 20 years, and considers him to be the Biology mentor for the Tarrant County College district. The four authors sincerely thank Mr. Nabors for his support, guidance and wisdom.

General College Biology Laboratory Safety Guidelines

Tarrant County College District

Student Name: _____ **Student ID#:** _____

You must complete the following for each lab section in which you are enrolled for each semester at TCCD. You will keep one copy. One copy will be kept by the instructor, and another, by a designated campus safety representative. TCCD copies are shredded after three years.

PRINT all answers.

Do you wear contacts? _____

List any allergies or medical problems that your lab instructor should be aware of.
_____
_____

List a contact person in case of emergency and a working phone number for him/her.
_____
_____

I have read, understand, and agree to follow these safety rules and procedures. I agree to abide by any additional instructions, verbal or written, provided by my instructor. I understand that if I do not comply with these safety rules, I may not participate in the lab activities and will forfeit any grades to be earned during that lab period. Continued noncompliance may result in my removal from the course.
_____
_____

Student Signature   Printed Name   Date

As your instructor, I assure you that I have read and understand these safety rules and procedures. I have agreed to enforce and follow them.

Instructor's Signature: _____

Printed Name and Date: _____

Course and Section Number: _____

Semester, and Year: _____

1 LABORATORY

The Scientific Method

Instructional Objectives

1. Be able to distinguish between hypothesis-driven science and discovery science.
2. Be able to identify the different parts of the scientific method and apply this to the experiment of the fermentation by yeast.
3. Be able to identify the control and the experimental factor in an experiment and apply this to the experiment of the fermentation by yeast.
4. Be able to distinguish between qualitative and quantitative observations.
5. Be able to distinguish between the independent and the dependent variable.
6. Be able to write the chemical equation for fermentation.
7. Be able to identify the products of fermentation by yeast and which one is measured.
8. Be able to plot a graph and to interpret data from a graph.

Materials

Warm water bath

24 fermentation tubes   Refrigerator

12 250-mL beakers   Red China marking pencil

1-liter container (molasses and yeast mixture at room temperature)

Part A. Science and the Scientific Method

Background Information

The scientific method is an attitude or a means to solve a scientific problem using critical thinking. It is an important way to formulate answers from questions. In science, questions are commonly referred to as scientific problems. Through the scientific method, evidence is collected under specific guidelines. These guidelines are part of the procedure of an experiment. The evidence that is collected may be published so that other scientists can review the evidence and verify that the evidence is accurate. If accurate, then the verified evidence may be incorporated into the body of scientific knowledge.

Though the scientific method is merely an attitude or a way of critical thinking that is used in science, one fact stands out and that is there is no particular sequence of steps present in the scientific method. However there are some core principles in the scientific method. The steps of the scientific method are as follows:

1. Problem
2. Hypothesis
3. Experiment
4. Conclusion
5. Paper
6. Theory

A problem is a question about a natural phenomenon. A natural phenomenon is described as an observable or measurable event that occurs under a specific set of conditions. First of all, the phenomenon must be repeatable. Miracles are not in the realm of science because they are not repeatable and may only occur once. Problems always start the statement with the terms *what* and *when,* but never start with *why* or *how*. An example of a problem is: What happens to the rate of respiration of a cold-blooded organism when its environmental temperature is changed?

A hypothesis is merely an educated guess or a plausible explanation. A hypothesis usually involves a cause–effect relationship, that is, an if–then relationship. In science the hypothesis must be testable. This means an experiment must be performed in order to test a particular hypothesis. An example of a hypothesis related to the problem is: If the temperature of the environment is warmed, the rate of respiration of a cold-blooded animal will increase.

An experiment must be designed to test a particular hypothesis. The two parts of an experiment are the control group and the variable (experimental factor). In each experiment there can be only one variable being tested at a time. The control is the group for which the conditions are not changed. The control is also called, in medical research, the *placebo group*. In a drug study, instead of receiving the drug, the placebo group will receive a sugar pill. Some patients will automatically feel better due to the thought that they are being cured. This is called the *placebo effect*. The conditions are exactly the same in an experiment except for the variable. The variable is what factor is being tested. There can be only one variable being tested in an experiment. The variable is also called the experimental group. From our hypothesis, one group of cold-blooded animals are kept in an environment at room temperature (22°C). This is the control. The other cold-blooded animals are kept in environments with different temperatures above room temperature. These are referred to as the experimental groups.

During the experiment observations are made. Observations are either qualitative or quantitative. A qualitative observation is made by using the senses of smell, taste, touch, or sight. These observations are not precise since the observer may not have the sensitivity that another observer may possess. When describing temperature, the terms, cold, cool, warm, or hot do not give a precise measurement. Quantitative observations require the use of instruments. An instrument is a device that extends the range of human senses or allows more precise determinations. For example, in our experiment of the rate of respiration of cold-blooded animals, if a thermometer were being used, the temperature can be measured in units of degrees Celsius rather using the terms cool, warm, or hot. Recorded quantitative observations may be arranged into groups of measurements known as data. Data can show the relationship of the variable to the measurements made. Data is presented in the form of tables and graphs.

The conclusion is a statement that is made at the end of the experiment that supports, refutes, or modifies the hypothesis. In the experiment of the effect of temperature on cold-blooded animals, if the data showed that the rate of respiration increased as the environmental temperature increased, then the conclusion proves that the hypothesis is correct. The experiment with its conclusion maybe submitted to a scholarly journal to be reviewed by other scientists. The experiment is repeated by other scientists to validate the experimental results.

After further experimentation, a theory may be formulated. A theory is an explanation of a phenomenon. It has been proven to be true with all available information. Scientific theories can be modified or can be proven to be false with the discovery of new information.

Part B. Fermentation

Most animals do not have the ability to regulate their own internal body temperatures. These animals that depend on the environmental temperature are referred to as cold blooded. With not being able to regulate the internal body temperature presents a problem for the survival of these cold blooded animals. Temperature is the measure of the average kinetic energy of the molecules in a system. The higher the temperature of a system means the higher the kinetic energy of the molecules. One factor that controls the rate of chemical reactions is the temperature of the system. If the temperature of a system increases by 10°C, the rate of the chemical reaction doubles. Metabolism is the sum of all of the chemical reaction occurring in an organism. In organisms which cannot regulate their internal body temperatures, their metabolic rate is determined by the environmental temperature. That is the reason why reptiles (snakes, turtles, and lizards) can be seen sunning themselves on a rock during cool mornings. These animals are absorbing infrared radiation so to increase their metabolism. As the rock absorbs the infrared radiation from the sun, the radiation is transformed into heat energy. The animal is absorbing the heat energy from the rock and the absorbed heat is raising the metabolic rate of the animal. The higher metabolic rate means that the animal can become more active. Of course, there is a limit to metabolism since most organisms, except for a few bacteria that live in hot springs where the water is at its boiling point (100°C), cannot survive temperature above 50°C. High temperatures denature or break down important proteins called enzymes. Enzymes are biological catalysts. A catalyst is a substance that speeds up a chemical reaction without itself being permanently altered. It is the enzymes that are responsible for allowing the chemical reactions in metabolism to occur at room temperature and above.

An important part of metabolism is respiration. The most common means of respiration occurs in living cells to produce energy from the sugar glucose ($C_6H_{12}O_6$). There are two forms of respiration: aerobic (*aerobic* means air) respiration and anaerobic (*an* means without; without air) respiration. Aerobic respiration occurs in the presence of air, more specifically in the presence of oxygen. Highly metabolic organisms, animals and plants, use aerobic respiration to produce their energy from glucose sugar. The following overall chemical equation represents the production of energy from glucose in the presence of oxygen. Anaerobic respiration is called fermentation. There are two types of fermentation: lactic acid fermentation and alcohol fermentation. If you have eaten yogurt or if you have experienced the burning feeling in your muscles while vigorously exercising, then you have experienced lactic acid fermentation. The other type of fermentation, alcoholic fermentation, results in the formation of ethyl alcohol (C_2H_5OH) and carbon dioxide (CO_2). Fermentation is used in the leavening (rising) of bread, the making of beer and wine, and the formation of the alternative fuel gasohol. In this experiment we will study the effects of temperature on fermentation by yeast.

Yeast is a unicellular fungus that may produce its energy in the presence of or in the absence of oxygen (O_2) gas. In fermentation, glucose ($C_6H_{12}O_6$) is broken down to produce ethyl alcohol (C_2H_5OH) and the gas carbon dioxide (CO_2). The reaction of alcoholic fermentation is as follows:

$$C_6H_{12}O_6 \rightarrow C_2H_5OH + CO_2$$

When you smell the odor of bread rising, what you smell is the ethyl alcohol being formed. The bread rising is due to the carbon dioxide (CO_2) being released into the dough.

Procedure

Students working in groups will follow these steps.

1. Obtain 3 clean fermentation tubes and label the tubes with a red China marking pencil make the following designations: your group symbol and one tube A, the other tube B, and finally the last tube C.

2. Obtain a mixture of baker's yeast and molasses solution at room temperature and fill the tubes with the mixture as instructed by your laboratory instructor. Make sure there is no air in the closed end of the fermentation tube. For best results the yeast-molasses mixture should be close to room temperature.

3. Place the fermentation tubes in the environmental conditions listed in Table 1.1 for approximately one hour.

TABLE 1.1   Fermentation Tubes

Fermentation Tube	Condition	Temperature (°C)
A	Refrigerator	4
B	Lab Table	22
C	Warm Water Bath	40

4. The gas formed is an indicator of the rate of fermentation. At the end of the time period, measure the amount of gas produced in the closed arm of the tube. **Mark the mixture levels in the fermentation tubes on Figure 1.1.**

5. Wash out the glassware with tap water after the experiment.

Tube A
(4° C)

Tube B
(23° C)

Tube C
(40° C)

FIGURE 1.1   Collection of CO_2 in Fermentation Tubes

Name: _Sweeney_

QUESTIONS

1. Write a hypothesis for this experiment:

   The tube in the warm water may have more bubbles

2. Which fermentation tube (A, B, or C) is the control?

   B

3. What is the variable in this experiment?

   temperature

4. Which substance formed by the yeast is measured to determine the rate of fermentation?

   CO_2

5. What is the role of sugar in this experiment?

   Catalyst

6. Comparing the mixture levels inside the fermentation tubes, does the data support or reject your hypothesis? Explain your answer.

```
  28              30.6              34              35.2
  32          5⟌153.0              37          5⟌176.0
  33            15↓                 36            15↓
  29            03                  34            26
  31            30                  38            25
 ____          ____               ____            10
  153                              176
```

Part C. Interpreting Data from Graphs

1. The Zebra Danio—*Brachydanio rerio*—is a tropical fish that is used in the study of genetics in development. The embryos are large enough to be seen through a microscope. In this experiment, the embryos are studied for the effect of temperature on their heart rate. Five fish were studied at the temperatures 0°C, 5°C, 10°C, 15°C, 25°C, 30°C, and 35°C, as shown in Table 1.2.

TABLE 1.2   Effect of Temperature (°C) on Heartbeat of Zebra Danio Embryos

Temperature (°C)	Number of Heartbeats/s					
	Fish A	Fish B	Fish C	Fish D	Fish E	Average
0	18	21	23	17	20	20
5	23	26	27	22	24	24
10	28	32	33	29	31	31
15	34	37	36	34	38	35
20	39	42	43	38	41	46
25	44	47	49	48	50	48
30	57	53	55	60	62	57
35	+	57	60	+	+	58

+ means that the embryo died during the experiment.

Find the average heart rate at each temperature and round off the average to the closest whole number.

Draw a graph of the average of temperature vs. heartbeat. Temperature is on the x-axis and number of heartbeats is on the y-axis. The x-axis is the independent variable. The independent variable is measurement that might influence the outcome. It is the condition that is changed in the experiment. It is also the presumed cause in an experiment. The dependent variable is the presumed effect in an experiment.

GRAPH 1.1   Effect of Temperature (°C) on Heartbeat of Zebra Danio Embryos

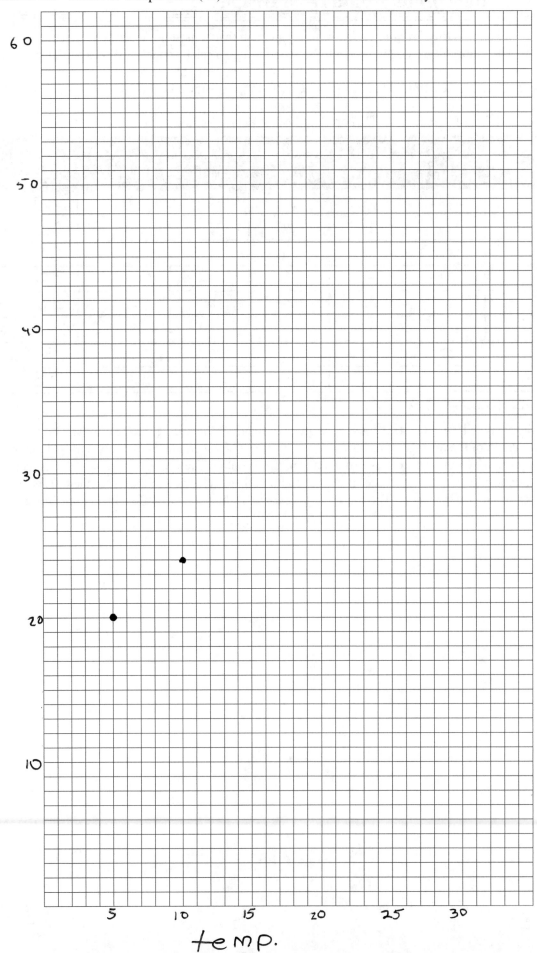

Name: _____

QUESTIONS

1.
  A. What is the relationship of temperature and heart rate in zebra fish?

  B. Are zebra fish warm-blooded or cold-blooded animals?

  C. Why did the most of the zebra fish die at 35°C?

2. A student used an instrument called a spectrophotometer in an experiment to determine the amount of light absorbed in the visible spectrum of light by chlorophyll A. The spectrophotometer is an instrument that separates light into individual wavelengths and measures the amount of light that is absorbed by pigments. The mixture used in this experiment contained chlorophyll that was extracted from spinach leaves.

   Sunlight (visible light) is actually composed of different wavelengths (colors) of light. A prism can bend light into its individual colors so that we may see them. When we see a rainbow, the water droplets in the air are acting like prisms that bend light into the bands of colors (red, orange, yellow, green, blue, indigo, and violet). Each color has its own specific wavelength. The color of visible light and its wavelength is found in Table 1.3, Color of Visible Light and Wavelength (nm).

TABLE 1.3   Color of Visible Light and Wavelength (nm)

Color	Wavelength (nm)
Red	750
Orange	600
Yellow	580
Green	500
Blue	475
Indigo	430
Violet	400

   Different wavelengths have different frequencies. The shorter the wavelength of light is inversely related to the higher the frequency of the light and the higher the frequency of the light—that is, the shorter the wavelength of light, the greater the energy of the photons of light. This leads to the problem of getting a sunburn on a cloudy summer's day. The clouds are blocking most of the longer wavelengths of light and allowing the blue and violet wavelengths, which have the greater energy.

The data from this experiment is shown in Table 1.4, Wavelength and Absorbance of Chlorophyll A

TABLE 1.4   Wavelength and Absorbance of Chlorophyll A

Color of Visible Light	Wavelength (nm)	Absorbance
	380	0.430
Violet	400	0.484
	420	0.620
Indigo	440	0.624
	460	0.468
Blue	480	0.300
	500	0.138
Green	520	0.098
	540	0.090
	560	0.092
Yellow	580	0.102
Orange	600	0.110
	620	0.121
	640	0.450
Red	660	0.540
	680	0.435
	700	0.053
	720	0.042

Plot these values as a curve on Graph 1.2, Absorption of Chlorophyll A

GRAPH 1.2   Absorption of Chlorophyll A

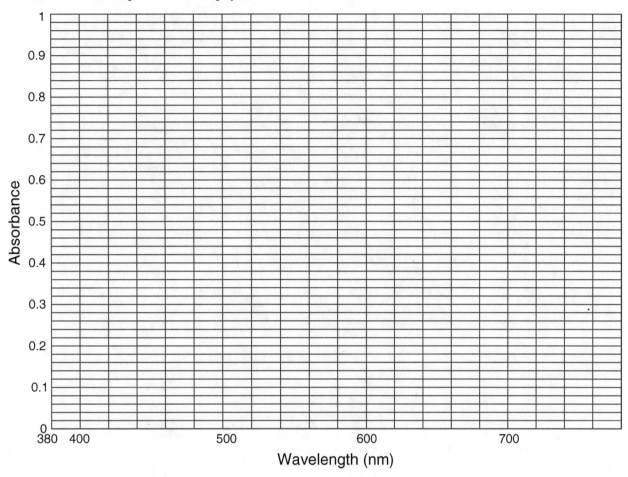

Name: _____

QUESTIONS

A. From the graph, which wavelength of light (color) is probably least essential to photosynthesis? Why?

B. From the graph, which wavelengths of light (colors) are absorbed the most?

C. What is your conclusion about the efficiency of the absorption of light in the process of photosynthesis?

2 LABORATORY

Factors Influencing Enzyme Activity

A. Instructional Objectives: After Completing this exercise the student should be able to:
1. Describe the relationship of molecular structure to enzyme function.
2. Describe the changes in enzymatic function by the influence by changes in the following:
   a. Enzymatic concentration
   b. Substrate concentration
   c. pH
   d. Ionic concentration
   e. Temperature
   f. Inhibitor
3. Distinguish between a dependent variable and an independent variable in an experiment.

B. Materials

3% hydrogen peroxide (H_2O_2)	Gas generators (12)
Yeast suspension in sucrose solution	Boiled yeast solution
100-mL graduated cylinders (12)	Stopwatch
Graduated Beral pipettes	

Yeast Suspension

Mix 7 g of dried yeast in 1000 mL of warm water and allow the yeast suspension to set for one hour at room temperature. Strain of the liquid using cheesecloth and discard the solids. Keep the liquid refrigerated until needed.

Stock Solutions

Hydrogen Peroxide (H_2O_2) per Section		
Concentration	H_2O_2	Distilled H_2O
3.0% H_2O_2	200	0
1.5% H_2O_2	500	500
0.75% H_2O_2	50	150
0.38% H_2O_2	25	175
0.0% H_2O_2	0	200

Buffer Solutions for the Semester	
pH	
3	982.3 mL 0.1 $HC_2H_3O_2$ + 17.7 mL 0.1 M $NaC_2H_3O_2$
5	295 mL 0.1 $HC_2H_3O_2$ + 705 mL 0.1 M $NaC_2H_3O_2$
7	600 ml 0.1 M KH_2PO_4 + 349.2 mL of 0.1 M NaOH
9	900 mL 0.025 M $Na_2B_4O_7$ • 10 H_2O (borax) + 82.8 mL of 0.1 M HCl
11	900 mL 0.05 M Na_2HPO_4 + 73.8 mL of 0.1 M NaOH

C. Background Information

Most chemical reactions inside a living cell do not occur spontaneously. The chemical reactions that occur with the cell are controlled by a group of proteins called enzymes. An enzyme acts as a catalyst. In chemical reactions the energy required for a specific chemical reaction to occur is called its activation energy. The activation energy for a chemical reaction is likened to bicycling up a hill. The higher the hill, the more energy the bicycler must exert to get over the peak. Of course, riding downhill requires no exertion of energy. The role of a catalyst is to lower the hill so that there is less energy required for the reaction to proceed. Lowering the hill means the reaction can more easily proceed. With a lower hill, the reaction rate increases. In essence, a catalyst is a substance that speeds up a chemical reaction with itself being permanently chemically altered.

It is important to remember that enzymes are three-dimensional molecules that have a specific shape. Each enzyme has a specific functional area where the substrate will fit, and that area is called the active site. The active site functions like a lock-and-key mechanism. The substrate is the reactant molecule that is chemically changed by the enzyme. The substrate fits into the active site of the enzyme. When this occurs, an enzyme–substrate complex is formed. When the complex is formed, the chemical bonds of the substrate are stressed so that they are broken and new chemical bonds are formed. This is how an enzyme lowers the activation energy by forming the enzyme–substrate complex.

Enzyme + substrate(s) → [enzyme–substrate complex] → enzyme + product(s)

When the product or products (are) formed, the enzyme is released to its original form and is able to repeat the process with other substrate molecules.

Enzymes are affected by different factors, which, in turn, affect the rate of the enzymatic reaction. These factors can change the shape of the enzyme molecule. The change in molecular shape of a protein is called denaturation. In this laboratory investigation, the effects of salt concentration, pH, temperature, and inhibitors will be studied.

a. Salt concentration: The amino acids in a protein have charged side chains that are attracted to one another to keep the shape of the protein. Ions also have either positive or negative charges. For example, common table salt, NaCl, is composed of a positive sodium ion (Na^+) and a negative chloride ion (Cl^-). When a salt such as NaCl dissolves in water—a sufficient amount to interfere with the side chains of the amino acids making up the enzyme—the enzyme molecule changes its shape and is said to be denatured. The normal amount of NaCl in human blood and cells is 0.9%.

b. pH: pH is a measurement of the amount of acid in a solution, that is, the amount of hydrogen ions (H^+) dissolved in a solution. pH is the –log [H+]. If a solution is 1×10^{-3} M H^+, the pH of the solution is 3. If a solution is 1×10^{-9} M H^+, the pH of the solution is 9. The scale runs from 0 to 14, with 0 being the most acidic solution and 14 being the most basic solution. If a solution has a pH of 7, then it is said to be neutral. pH + pOH = 14.00. If the pH of a solution is 3, then the pOH is 11. This means there are more 100,000,000 times more H^+ ions than OH^- ions in the solution. If the pH of a solution is 9, then pOH is 5. This means there are 10,000 times more OH^- ions than H^+ ions in the solution.

Each amino acid contains an –NH3+ group and –COO$^-$ group. When an amino acid is placed in an acidic environment with an excess of H+ ions in an acidic solution, the H+ ions react with the –COO$^-$ group to form COOH. When an amino acid is placed in a basic environment, there are few H^+ ions but an excess of OH^- ions; the –NH_3^+ reacts with –OH^- to form H_2O. In both cases, the amino acids change their charge and the enzyme molecule is denatured.

c. Temperature: Generally for every 10°C increase within a system, the rate of a chemical reaction doubles. Temperature is the measure of the average kinetic energy of the molecules within a system. As the temperature increases, the kinetic energy increases. The faster the molecules are moving, the greater the rate of collisions between the molecules. There is an optimal range for enzymes to function. If the temperature of the system exceeds the optimal of the enzyme, the molecular shape of the enzyme becomes distorted. As a the result, the active site of the enzyme changes shape where the substrate molecules cannot lock in. Most protein molecules become denatured around 40 to 50°C.

Hydrogen peroxide (H_2O_2) is a poisonous by-product of aerobic respiration in the cell. Catalase is an enzyme that decomposes hydrogen peroxide into water and oxygen gas.

$$2 H_2O_2 \rightarrow H_2O + O_2$$

Yeast cells that have been suspended in a sucrose solution for an hour at room temperature are the source of catalase in this laboratory investigation.

D. Procedure

Students will be working in groups to complete the lab.

Procedure A: The Effect of Substrate Concentration on Enzyme Activity

1. Completely fill a 100-mL graduated cylinder with water and invert it into a large beaker. Make sure there are no air bubbles in the graduated cylinder.

2. Remove the top from the gas reaction vessel. As prescribed by your laboratory instructor, use a Beral-style pipette and add to the reaction vessel the specified volume of hydrogen peroxide and concentration to the reaction vessel. Stir your source of catalase (yeast extract). Also add, as prescribed by your laboratory instructor, the specified amount of catalase. Seal the reaction vessel as instructed and quickly pinch the plastic tubing. Position the tube to take the gas evolved enters the graduated cylinder.

Concentration	mL of H_2O_2	mL of Catalase
3.0% H_2O_2		
1.5% H_2O_2		
0.75% H_2O_2		
0.38% H_2O_2		
0.0% H_2O_2		

3. Record the volume of oxygen evolved every 30 intervals for 5 minutes or less if bubbling ceases.

Use Table 2.1 and plot the data points for all concentrations of H_2O_2 on Graph 2.1, where the y-axis is volume (mL) of oxygen gas evolved and the x-axis is time (sec).

Procedure B: The Effect of pH on Enzyme Activity

1. Completely fill a 100-mL graduated cylinder with water and invert it into a large beaker. Make sure there are no air bubbles in the graduated cylinder.

2. Remove the top from the gas reaction vessel. As prescribed by your laboratory instructor, use a Beral-style pipette and add to the reaction vessel the specified volume of 1.5% hydrogen peroxide and volume of buffer to the reaction vessel. Stir your source of catalase (yeast extract). Also add, as prescribed by your laboratory instructor, the specified amount of catalase. Seal the reaction vessel as instructed and quickly pinch the plastic tubing. Position the tube to take the gas evolved enters the graduated cylinder.

pH	mL of Buffer	mL of 1.5% H_2O_2	mL of Yeast Extract
3			
5			
7			
9			
11			

3. In Table 2.2, record the volume of oxygen evolved every 30 intervals for 5 minutes or less if bubbling ceases; plot the data points for all pH conditions on Graph 2.2, where the y-axis is the volume (mL) of oxygen gas evolved and the x-axis is time (sec).

Procedure C: The Effect of Salinity on Enzyme Activity

1. Completely fill a 100-mL graduated cylinder with water and invert it into a large beaker. Make sure there are no air bubbles in the graduated cylinder.

2. Remove the top from the gas reaction vessel. As prescribed by your laboratory instructor, use a Beral-style pipette and add to the reaction vessel the specified volume of 1.5% hydrogen peroxide and volume of saline solution to the reaction vessel. Stir your source of catalase (yeast extract). Also add, as prescribed by your laboratory instructor, the specified amount of catalase. Seal the reaction vessel as instructed and quickly pinch the plastic tubing. Position the tube to take the gas evolved enters the graduated cylinder.

Concentration Saline Solution	mL of 1.5% H_2O_2	mL of Catalase
10.0% NaCl		
2.0% NaCl		
0% H_2O_2		

In Table 2.3, record the volume of oxygen evolved every 30 intervals for 5 minutes or less if bubbling ceases; plot the data points for all salinity conditions on Graph 2.3, where the y-axis is volume (mL) of oxygen gas evolved and the x-axis is time (sec).

Procedure D: The Effect of Temperature on Enzyme Activity

1. Completely fill a 100-mL graduated cylinder with water and invert it into a large beaker. Make sure there are no air bubbles in the graduated cylinder.

2. Remove the top from the gas reaction vessel. As prescribed by your laboratory instructor, use a Beral-style pipette and add to the reaction vessel the specified volume of 3.0% hydrogen peroxide. Stir your source of catalase (yeast extract). Also add, as prescribed by your laboratory instructor, the specified amount of catalase. An ice bath is required for 5°C. Room temperature is assumed to be 22°C.

Temperature (°C)	Volume of 3% H_2O_2	Volume of Catalase
5		
22		
37		
Boiled yeast solution		

3. Place the reaction vessel in the water bath. Allow four minutes for the system to adjust to the new temperature. Position the tube so that the gas evolved enters the graduated cylinder.

4. In Table 2.4, record the volume of oxygen evolved every 30 intervals for 5 minutes or less if bubbling ceases; plot the data points for all salinity conditions on Graph 2.4, where the *y*-axis is volume (mL) of oxygen gas evolved and the *x*-axis is time (sec).

TABLE 2.1   The Effect of Substrate Concentration on Enzyme Activity

Time (sec)	Concentration of Hydrogen Peroxide (H_2O_2)				
	3.0% H_2O_2	1.5% H_2O_2	0.75% H_2O_2	0.38% H_2O_2	0.0% H_2O_2
0					
30					
60					
90					
120					
180					
240					
300					

TABLE 2.2   The Effect of pH on Enzyme Activity

Time (sec)	pH				
	3	5	7	9	11
0					
30					
60					
90					
120					
180					
240					
300					

TABLE 2.3   The Effect of Salinity on Enzyme Activity

Time (sec)	Salinity		
	10% NaCl	2% NaCl	0% NaCl
0			
30			
60			
90			
120			
180			
240			
300			

TABLE 2.4   The Effect of Temperature on Enzyme Activity

Time (sec)	Temperature			
	5°C	22°C	37°C	100°C
0				
30				
60				
90				
120				
180				
240				
300				

GRAPH 2.1   The Effect of Substrate Concentration on Enzyme Activity

GRAPH 2.2   The Effect of pH on Enzyme Activity

GRAPH 2.3   The Effect of Salinity on Enzyme Activity

TABLE 2.4   The Effect of Temperature on Enzyme Activity

Name: _____

QUESTIONS

1. What was the source of catalase in this experiment?

2. What is an enzyme?

3. What is a substrate?

4. Write the equation for the decomposition of hydrogen peroxide.

5. What is the gas that you see bubbling up?

6. Why is it important for cells to have catalase present?

7. What was the effect of substrate concentration on enzyme activity?

8. What was the effect of pH on enzyme activity?

9. From your graph what was the optimal pH for the enzyme activity of catalase?

10. What happened to the rate when the pH was above or below the optimal range?

11. When the rate was 0 under certain pH conditions, why was the enzyme not functional?

12. What was the effect of temperature on enzyme activity?

13. From your graph what was the optimal temperature for catalase?

14. At what temperature was the rate 0?

15. Why was the enzyme not functional at that temperature?

LABORATORY

Care and Use of the Microscope

A. Instructional Objectives

1. Identify and define all parts of the microscope, and understand the function of each.

   a. Ocular lens, body tube, revolving nosepiece, objective lenses, light diaphragm lever, light source, base, coarse adjustment knob, fine adjustment knob, stage, stage clips, and arm.

   Please consult Table 3.1.

TABLE 3.1 Microscope Parts and Their Functions

Ocular lens	Lens nearest the eye
Rotating nosepiece	Rotates the various objective lenses
Objective lenses	Contains lenses of various magnification
Pointer	Found in ocular lens, usually right
Base	Bears the weight of the microscope
Mechanical stage	Supports the slides
Coarse adjustment knob	Moves stage up and down, focus
Fine adjustment knob	Permits exact focusing
Condenser	Condenses beam of light onto slide
Iris diaphragm lever	Regulates amount of light for contrast
Arm	Supports body of microscope while carrying

2. Identify and define the ocular and objective lenses, state their magnification, and calculate the total magnification of any lens combination.

3. Determine the diameter of the field of vision using each objective lens.
4. Use the field of vision to estimate an object's size. Define and understand the concept of working distance.
5. Set up, carry, and use the microscope properly.
6. Determine reasons for not finding an object on a slide.

B. Introduction

The binocular compound light microscope is a useful tool in the study of biology. The microscope will be used to view objects too small to be seen with the unaided eye. The microscope has three useful features that allow scientists to see objects on a slide. They are magnification, resolution (resolving power), and contrast. The microscope that the student uses in this lab magnifies objects up to a thousand times, and the resolving power of the microscope allows the student to see the object clearly. Contrast refers to the object seen against a lighter or darker background, and can be controlled by opening or closing the diaphragm on the microscope. The microscope is an expensive instrument, and requires accurate use and care by the student.

C. Materials
   1. Per student: 1 microscope, 1 prepared slide of 3 colored threads, 1 prepared slide of letter "e," lens paper, 1 plastic metric ruler.
   2. Per laboratory: Protoslo® (methyl cellulose), used to slow down the movement of live organisms, staining solutions, glass slides, coverslips, and prepared stained slides.

D. Methods
   1. Transporting the microscope: Always carry the microscope by its arm in an upright position, with free hand beneath the base. Always carry the microscope alone. Do not pick up slide box until the microscope has been removed from the cabinet, and placed on student lab table.
   2. Learn the microscope parts: Compare the microscope to Figure 3.1 of this exercise. Learn the name, location, and function of all parts listed in the figure.
   3. Setting up the microscope for use:
      a. Clean ocular and objective lenses, using the lens paper.
      b. Place scanning objective (x 4) in position, directly over the light source. Make sure the objective "clicks" in position, so that it cannot easily be jolted. This ensures that the diameter of the objective is open.
      c. Turn on the substage light.
      d. Place the slide you wish to observe on the stage with the coverslip up.
      e. Raise the stage as far as possible with the coarse adjustment knob.
      f. Looking through the ocular lens, slowly raise the stage with the coarse adjustment until the specimen is viewed clearly.
      g. Sharpen the focus with the fine adjustment knob.
      h. To use the low-power objective, turn the revolving nosepiece until the low-power objective clicks in place.
      i. Locate, center, and focus the specimen in the field of view.
      j. Carefully rotate the nosepiece until the high-power objective is in place.
      k. Using the fine adjustment, sharpen the image.
      l. Never use the coarse adjustment knob with the high-power objective. If focusing problems occur, go back to scanning lens, refocus, and recenter.
   4. Reasons for indistinct image: The objective lens may be dirty or wet, an eyelash is on the slide, grease from the fingers is on the slide or lens, the specimen may be too thick, the specimen may be out of the field of view.
   5. Physical properties of microscope specimen: While observing the prepared slides and the wet mount slides, answer the questions listed at the end of the exercise.
      a. Depth of focus: this is referring to being able to see in the third dimension. Use the colored thread slide for this. Focus the center of the colored threads with the coarse adjustment knob, then adjust the contrast with the iris diaphragm lever. You should be able to see that each colored thread

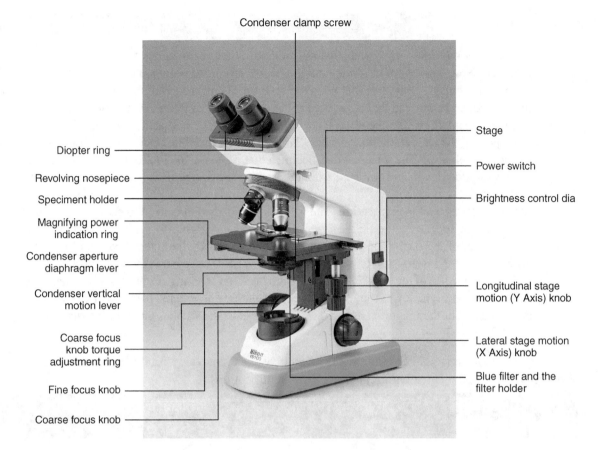

Condenser clamp screw

Diopter ring

Revolving nosepiece

Speciment holder

Magnifying power
indication ring

Condenser aperture
diaphragm lever

Condenser vertical
motion lever

Coarse focus
knob torque
adjustment ring

Fine focus knob

Coarse focus knob

Stage

Power switch

Brightness control dia

Longitudinal stage
motion (Y Axis) knob

Lateral stage motion
(X Axis) knob

Blue filter and the
filter holder

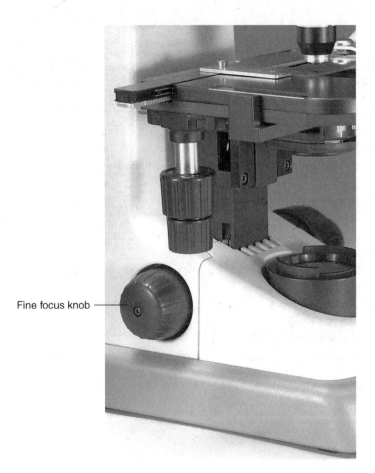

Fine focus knob

FIGURE 3.1   Courtesy Nikon Instruments, Inc.

is made up of multiple tiny threads that are braided. When the image is focused, deliberately blur the image. Now start again to focus, and notice which thread comes into focus first. This is the most inferior thread, or the one that is on the bottom. The next thread to come into focus is in the middle, and the last thread that is focused is the one on top, or the most superior thread.

b. Specimen orientation: Use the slide labeled "letter e." First, look at the orientation of the letter before putting the slide in the microscope. Then, examine the slide under the scanning lens. What is its orientation now? The image of the specimen that is seen while looking through the ocular lens is always upside down, and reversed from the specimen as seen on the slide when it is not in the microscope.

c. Working distance and magnification: The following information is used to estimate the size of the specimen as it appears in the field of view. First, total magnification is calculated by taking the magnification of the ocular lens (x10) and multiplying that by the magnification of the objective lens in place. Working distance is defined as the physical space between the bottom of the objective lens and the top of the slide. The relationship between working distance and the magnification is an inverse one. The lower the magnification of the objective lens, the greater the working distance. The higher the magnification of the objective lens, the smaller the working distance. Please consult Table 3.2.

TABLE 3.2 Comparison of the Relative Diameters of Fields of View, Total Magnification, and Working Distances for Three Different Objective Magnifications

Total Magnification with 10× Oculars	Working Distance	Diameter of Field of View
40× with scanning lens	With 4× scanning lens, 25 mm	With 4× scanning lens, 4.5 mm
100× with low-power lens	With 10× low-power lens, 8.3 mm	With 10× low-power lens, 1.8 mm
400× with high-power lens	With 40× high-power lens, 0.5 mm	With 40× high-power lens, 0.45 mm

6. Preparation of fresh wet mounts: Use an eyedropper, and place 1 drop of the culture water in the center of a clean glass slide. Always use a coverslip over the drop of liquid. Carefully place the coverslip at the edge of the liquid, and allow it to fall slowly back onto the liquid. This allows most air bubbles to escape. Make wet mount preparations of all cultures available for study in the lab. Some of the faster moving organisms need to be slowed down by using Protoslo®. If necessary, use a drop of stain to see the specimen better.

7. Examination of commercially prepared slides: Examine the prepared slides available for study.

8. Storing the microscope:
   a. Remove all slides from the stage, and place in box of prepared slides, or dispose of the slide properly.
   b. Clean all lenses using lens paper.
   c. Turn the scanning objective lens in place over the light source.
   d. Lower the body tube all the way down.
   e. Wrap electric cord around the base.
   f. Return the scope to its designated area, carrying it with both hands.

Name: _____

QUESTIONS

1. What color thread was on top in the prepared slide you examined?

2. What color thread was on bottom?

3. Draw the letter "e" as it appeared when viewed through the ocular lens.

4. If you move the slide to the right on the stage, in what direction does it appear to move when viewed through the microscope?

5. If you move the slide toward you, in what direction does it appear to move when viewed through the microscope?

6. What is the total magnification of the specimen as seen through the scanning lens objective? Low-power lens? High-power lens?

7. What microscope part regulates the amount of light that passes through the stage opening?

8. Which objective is always used first when beginning the search for an object on a slide?

9. Is it permissible to adjust the stage up while looking through the microscope? Why or why not?

diffusion high to low concentration

4 LABORATORY

Movement Across Membranes

A. Instructional Objectives: After completing this exercise, the student should be able to:

1. Describe Brownian movement.
2. Define, give examples of, and answer questions about diffusion, osmosis, dialysis, and filtration.
3. Explain the effects of temperature and molecular size on rates of diffusion.
4. Describe a selectively permeable membrane.
5. Define the terms isotonic, hypotonic, and hypertonic. Describe what happens when cells are placed in solutions of these three types.
6. Describe the significant factors that account for different concentrations of glucose and sodium chloride being isotonic to one another.
7. Explain how the rate of osmosis is changed for the concentration of different solutes.
8. Describe the chemical reagents used to test for the presence of sugar and starch.
9. Distinguish between electrolytes and nonelectrolytes.

B. Introduction

The cell is the basic functional and structural unit of life. All living things are made up of cells. The cell has an outer boundary called the plasma membrane, which protects the contents of the cell and acts as a gatekeeper for substances entering and leaving the cell. Some substances, such as small, nonpolar molecules, cross the plasma membrane easily. Examples include the respiratory gases, such as oxygen and carbon dioxide. Others, such as large molecular weight molecules or charged particles, cross the membrane more slowly, or with difficulty. Examples are proteins and starches. The plasma membrane is described as being selectively permeable (or semipermeable), because of its ability to regulate the molecules crossing the plasma membrane to enter and leave the cell. These processes that are responsible for the movement of materials across cell membranes are the topics of this exercise. Diffusion is defined as the movement of solids across a selectively permeable membrane from an area of high concentration to an area of low concentration. Osmosis is defined as the movement of liquids (in most cases, this refers to water) across a selectively permeable barrier, or membrane, from an area of high liquid concentration to one of low liquid concentration. Dialysis is defined as the separation of small dissolved particles from larger particles by means of diffusion across a selectively permeable membrane.

C. Materials (enough for six groups of four students each)

Textbook and lab manual

Beakers: 1000, 600, 400, 200 mL—6 each

String, and dialysis tubing

Ring stands and rings

Food coloring, assorted

Potassium permanganate crystals and methylene blue crystals

Sugar, starch, and filter paper

Graduated cylinders: 10 mL—4 each

Iodine solution

One bottle of India ink and small concave glass to see it (Students may also use glass slides.)

Test tubes, test tube racks, and test tube holders

Labels for test tubes

Lightbulb assembly

Glass marking pencils

Water bath, boiling

Water bath, with ice cubes

Hot plates

Distilled water

Agar plates

Prepared solutions of NaCl, NaOH, HCl, glucose, starch—1 each, for the entire lab

D. Procedures

The students will group together in six groups of four students and perform the exercise in groups. Each student is responsible for answering all questions.

1. Brownian movement is defined as erratic, nondirectional movement, observed by the microscope in suspensions of light matter that result from the jostling or bumping of larger particles by the molecules in the suspending medium, which are in continuous motion. To see this, place a drop of diluted India ink (charcoal particles) on a slide and cover it with a cover slip. Adjust and observe under high power magnification. Describe and explain what is seen in the space below.

_____

2. Diffusion and particle size: Particle size is estimated by knowing the molecular weight of the substance under study. The smaller the substance, the more affected it is by the kinetic force that causes diffusion. Diffusion is defined as the movement of molecules down a concentration gradient across a selectively permeable membrane. This means that solids in an aqueous solution, known as the solutes, will move from a region where they are more highly concentrated to a region where they are less concentrated. Take an agar plate and the two dyes potassium permanganate (molecular weight [MW] 158) and methylene blue (MW 320). Place small amounts of each dye in the center of different agar plates. Allow to sit undisturbed for an hour, and watch the diffusion of both dyes, comparing the difference in diffusion rate. Record results and an explanation below.

3. Diffusion and the effect of temperature: There is a direct relationship between heat and the rate of molecular motion, and thus, the rate of diffusion for a solute. Prepare two 1000-mL beakers of water. The water in one beaker should be near the freezing point (use the ice cubes to reduce the temperature), and the water in the second beaker should be near the boiling point. Place the second beaker of water on the hot plate, and heat it until it is almost boiling. Place identical drops of food coloring into the two beakers, and watch for several seconds. Record the appearance of the two solutions.

Color in the cold water: _____ blue _____

Color in the hot water: _____ red _____

Explain the effect of temperature on the rate of diffusion: _In higher temps,_ _the molecules are moving, causing the_ _dye to mox as well_

4. Simultaneous osmosis and diffusion: In the lab, dialysis membrane is a fake membrane used to mimic the action of the cell's plasma membrane. It has a microscopic pore size that allows only small molecular weight molecules to pass through. Large molecular weight molecules will not move through the microscopic pores. Water has a molecular weight of 18. Starch has a molecular weight exceeding 30,000. The dialysis tubing is defined as "differentially permeable," meaning it will allow certain substances to pass through it but not others.

Obtain a cut section of the tubing that has been soaked in distilled water. Tie off one end of the tubing with string to form a leak proof bag. Open the other end, and fill the tubing with a 1% starch solution. Tie off the second end, weigh the bag, and record the weight in grams.

What is the weight? _____ 2. 6 _____

Place the bag in a beaker filled halfway with distilled water. Note the time. Leave the setup in place for one hour.

Get two test tubes, and fill them halfway with distilled water (test tube #1) and 1% starch solution (test tube #2). Place them in a test tube rack. Obtain iodine, which is the reagent for starch, and add 2 drops of the iodine solution to both test tubes. A positive test for the presence of starch is a black inky color. A negative test for its presence is diluted iodine. Clean test tubes, and use them again in an hour.

After one hour, take the bag out of the beaker of water and weigh it again.

What is the weight? _____ 4. 6 _____

Has osmosis occurred? _____

How do you know? _____

Use the same two test tubes as before. Pour the water from the beaker in test tube #1, and add iodine to it. What is the test? _____

Cut open one end of the bag, and pour the starch contents into test tube #2. Add iodine and observe the test for the presence or absence of starch. _____

Did starch diffuse from the bag into the beaker after one hour? _____

Why or why not? _____

5. An example of tonicity in plant cells: observe Figure 4.1, "Hyper/Hypotonicity"

The concept shown in Figure 4.1 is one of a plant cell submerged in two different aqueous solutions. One solution is distilled water, and the other one is a 10% salt solution. The term *tonicity* means the ability of a solution to change the shape of a living cell. The only way this can happen is if water either leaves or enters the cell so that there is a net efflux or influx of water. The normal solution in a living cell is 0.9% NaCl and 99.1% water. The two solutions above do not represent the normal solution in living cells.

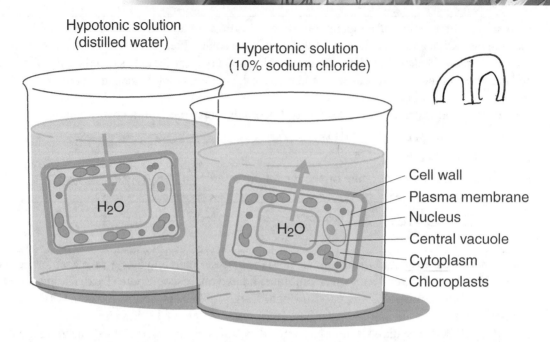

Hypotonic solution
(distilled water)

Hypertonic solution
(10% sodium chloride)

H_2O

H_2O

Cell wall
Plasma membrane
Nucleus
Central vacuole
Cytoplasm
Chloroplasts

A. Net flow of water into cell   B. Net flow of water out of cell

FIGURE 4.1   Initial Conditions for Osmosis and Diffusion Experiment, Showing Composition of Solutions Inside and Outside Dialysis Bag

From *Biological Investigations: From, Function Diversity and Process,* 6th edition by Warren Dolphin, © 2002. Reprinted by permission of The McGraw-Hill Companies.

The hypotonic solution is one with a solute concentration less than 0.9% NaCl. The best way to use this kind of solution in the lab is to use distilled water, which is 100% water. It contains no solutes. Water always moves from where there is more of it to where there is less. Thus, water moves into the cell (follow the blue arrow), and the cell will start to swell. A plant cell, having a cell wall in addition to a plasma membrane, will permit the swelling and will contain it. This allows a plant to grow, and the condition is called turgid.

The hypertonic solution is one with a solute concentration greater than 0.9% NaCl, and the solution above is shown at 10% NaCl, and thus, 90% water. As water follows its gradient, it leaves the cell and the plasma membrane pulls away from the cell wall. This condition in plants is called plasmolyzed.

6. Detection of electrolytes: Note that this procedure can be done as a demonstration by the lab instructor for the entire class. Students are responsible for obtaining the results. Ionic solutions such as acids, bases, and salts are called electrolytes in body fluids because they are electrically charged and will conduct an electric current. Ionic substances break up or dissociate in water, resulting in separate ions, which have either a positive or negative charge. Conductivity can be tested by the use of a lightbulb assembly. Plug the assembly into an electrical outlet, and notice the bulb does not glow because the flow of electricity is interrupted by the gap between the electrodes. Do not touch the electrodes. Connecting the electrodes by a conductor of electricity completes the circuit and the bulb lights up. An electrolyte solution will complete the circuit as the charged ions move between the electrodes. Use two lightbulb assemblies with different size bulbs (200 watts and 15 watts): this makes it easier to distinguish between bright and dim light.

Saturated solutions of glucose, starch, sodium chloride, hydrochloric acid, and sodium hydroxide have been prepared. These solutions and water are to be tested for the presence of ions. Raise each beaker of fluid until the electrodes are submerged but the bottom of the beaker is not touched. Record the results below:

Solution Tested	Results (Bright light, Dim light, No light)
Distilled water	*None*
Tap water	*Dim*
Starch	*None*
Dextrose	*None*
NaCl (salt)	*Both*
HCl (acid)	*Both*
NaOH (base)	*Both*

7. Match the process in column A with the description in column B.

Column A	Column B
A. Dialysis	*E* 1. Water flowing over a dam
B. Diffusion	*D* 2. Cells bursting in distilled water
C. Filtration	*D* 3. Meat being dried and preserved with salt
D. Osmosis	*B* 4. Perfume reaching nose of 1 person
E. None of these	*X* Urea diffusing from blood into water
	C 6. Milk at dairy poured through cotton pad for cleaning
	X Water forced from blood into tissues by blood pressure
	B 8. Exchange of O_2 and CO_2 between blood and air sacs in lungs
	B 9. Sugar cubes sweetening tea with no stirring

8. Fill in the following blanks:

_____1. Name given to the movement of particles due to molecular motion.

_____2. Diffusion, osmosis, dialysis, and filtration are all examples of (physical or physiological) processes.

_____3. Which term is the most inclusive: osmosis, diffusion, or dialysis?

Water 4. As osmosis occurs in the body, what substance is crossing a selectively permeable membrane?

_____ *X* When other factors are held constant, which substance—glucose or salt—should diffuse faster?

_____6. In osmosis, net movement of water is into a (hypertonic or hypotonic) solution.

_____ 7. What material is used to test for starch?

OSMOSIS 8. The membranous bag used to demonstrate dialysis swells and may even burst open. What is the obvious explanation (one word)?

filtration 9. Which of these processes is not a special type of diffusion: osmosis, filtration, or dialysis?

electrolytes 10. What term is used for substances that ionize in water?

_____ 11. What three types of compounds are electrolytes?

acids
bases
salt

5 LABORATORY

Cell Structure and Function

A. Instructional Objectives: After completing this exercise, the student should be able to:

1. Define all bold-faced terms.
2. Identify on a cell model and/or diagram and list the major functions of the plasma membrane and cytoplasm.
3. Identify on a cell model and/or diagram and list the major functions of the following cell organelles:

Nucleus	**Ribosomes**
Smooth endoplasmic reticulum	**Golgi apparatus**
	Lysosome
Rough endoplasmic reticulum	**Mitochondria**
	Centrioles
Vacuole	**Cell wall**
Chloroplast	

4. Identify on a cell model and/or diagram and list the major functions of the following parts of a nucleus:

   Nuclear envelope

   Nuclear pores

   DNA

   Nucleolus

5. Be able to identify the following organisms under the microscope and classify what type of cell they are:
   - i. Cheek cell
   - ii. *Elodea* (Needs to be italicized)
   - iii. Mixed protozoa
   - iv. *Penicillin*
   - v. *Rhizopus*
   - vi. *Anabaena*
   - vii. Bacteria (coccus, bacillus, spirilla)
6. Be able to describe and perform the steps of a wet mount preparation.

B. Introduction

The **cell** is the structural and functional unit of all living things and is very complex. The two major types of cells are the eukaryotic cell and prokaryotic cell. **Eukaryotic cells**, unlike prokaryotic cells, are complex and contain membrane-bound organelles. Eukaryotic cells are very diverse and include four types: animal, plant, protist, and fungi. **Prokaryotic cells** include many types of bacteria, cyanobacteria, and archaea.

Any cell, eukaryotic or prokaryotic, has a specific size, shape, and internal composition that reflects the type of organism it is found in. Nonetheless, all cells have common anatomical features and carry out certain functions. This exercise focuses on the structures and **organelles** that carry out those metabolic functions for the "typical" cell. A "typical" animal cell and plant cell will be used to study these structures and organelles.

C. Materials

Three-dimensional model of the animal cell and plant cell

Compound light microscope

Per student: 2 glass slides, 2 coverslips, methylene blue, toothpick, normal saline

Elodea plant

Mixed protozoan slide

Fungus (*Penicillin, Rhizopus*)

Three types of bacteria (coccus, bacillus, spirilla) slide

Cyanobacteria (*Anabaena*)

D. Methods

1. Cell Structures
    Chart 5.1 contains a list of structures/organelles found in a typical plant or animal cell. Identify each of these on an animal or plant cell model and know their function.
    Use Figures 5.1 and 5.2 as guides for identifying the structures.

2. Wet Mount Preparation
    a. Animal Cell. Obtain a glass microscope slide. With a clean toothpick, gently scrape the inside of your cheek. Then wipe the scraping onto the glass slide. Since it is difficult to see the scraping, roll the toothpick around on the slide several times. Then place a drop of normal saline solution over the scraping. Place the coverslip at a 45 degree angle to the edge of the specimen on the slide. Slowly lower the coverslip over the specimen, being careful not to trap any air bubbles underneath. In order to see the specimen, place a drop of methylene blue on the edge of the coverslip. Then place a paper towel at the opposite edge of the coverslip to draw the methylene blue under the coverslip. Notice the following structures in each cell:
        i. Cytoplasm
        ii. Plasma membrane
        iii. Nucleus
    b. Draw the cheek cell

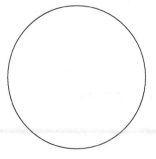

c. Plant Cell. Obtain a glass microscope slide. Place a leaf of *Elodea* in the middle of the slide. Then place a drop of normal saline solution over the leaf. Place the coverslip at a 45 degree angle to the edge of the specimen on the slide. Slowly lower the coverslip over the specimen, being careful not to trap any air bubbles underneath. Notice the following structures in each cell:

   i. Cytoplasm
   ii. Plasma membrane
   iii. Nucleus
   iv. Chloroplasts

d. Draw the plant cell

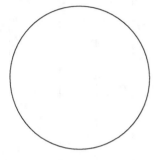

CHART 5.1   Cell Structure/Organelle Functions

Organelle/Structure	Function
Plasma Membrane	Regulates/allows passage of molecules in and out of cell
Cell Wall	Structural support of cell Found in plant cells, NOT animal cells
Cytoplasm	Water-based medium that suspends organelles and structures within the cell
Nucleus	Directs all activities within the cell Contains DNA, the genetic material of the cell/organism in the form of chromosomes and/or chromatin Contains the nucleolus, the site of ribosome synthesis
Endoplasmic Reticulum (ER)	System of membranous channels/saccules physically continuous with nuclear envelope Rough ER-studded with ribosomes, storage and transport of proteins Smooth ER-steroid and lipid synthesis, lipid metabolism
Cytoskeleton	Cell movement-microfilaments, intermediate filaments and microtubules Organelle movement & suspension Cell division
Vacuole	Storage of wastes, water, ions, etc. . . .
Chloroplast	Photosynthesis, conversion of solar energy to chemical energy Have their own DNA and ribosomes Storage of photosynthetic products Found in photosynthetic organisms (e.g., plant cells), NOT animal cells
Ribosomes	Site of protein synthesis Can be free or bound to endoplasmic reticulum (rough ER)
Golgi Apparatus	Processes material synthesized by ER Packages material and provides "address label" for transport out of/within the cell
Centrioles	Cell division Control cilia and flagella movement Found in animal cells, NOT plant cells

CHART 5.1   Cell Structure/Organelle Functions—Cont'd

Lysosome	Contains hydrolytic enzymes for digestion
	Digest material engulfed by cell
	Digest and recycle damaged organelles
Mitochondria	Production of ATP, the energy supply for the cell
	Glucose catabolism
	Have their own DNA and ribosomes
Nuclear Envelope (Membrane)	Regulates/allows passage of molecules in and out of nucleus

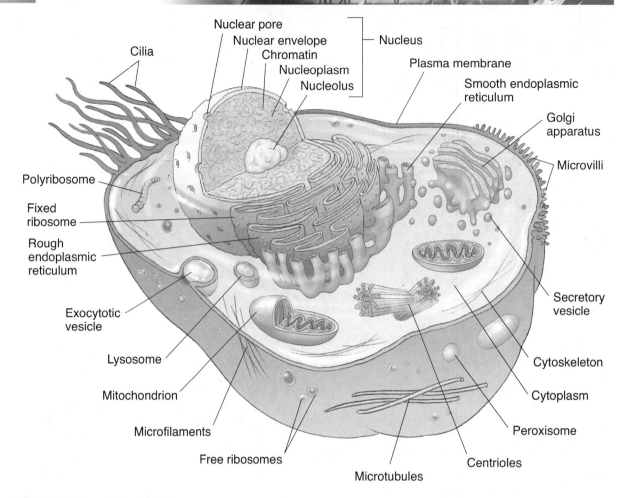

FIGURE 5.1   Animal Cell

Copyright © Kendall/Hunt Publishing Company.

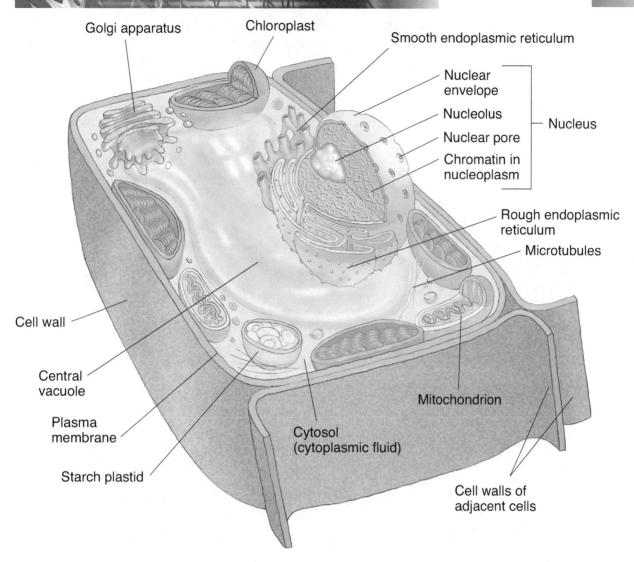

FIGURE 5.2 Plant Cell
Copyright © Kendall/Hunt Publishing Company.

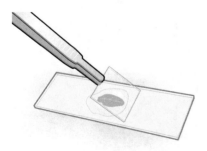

FIGURE 5.3 Wet Mount Preparation
Copyright © Kendall/Hunt Publishing Company.

3. View the following cells under the microscope.

a. Mixed protozoa

b. *Penicillin*

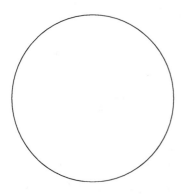
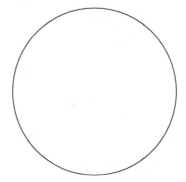

c. *Rhizopus* (black bread mold)

d. *Anabaena*

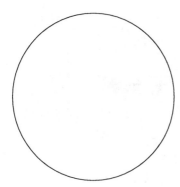
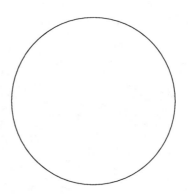

e. Bacteria (coccus, bacillus, spirilla)

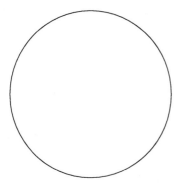

Name: _____

QUESTIONS

1. Match each of the cell organelles/structures with its function.

Smooth ER	_____	a. production of ATP
Golgi apparatus	_____	b. site of protein synthesis
Ribosomes	_____	c. package of proteins
Nucleus	_____	d. contains most DNA
Centrioles	_____	e. digests/recycles cell parts
Mitochondria	_____	f. store wastes, water, . . .
Chloroplast	_____	g. site of photosynthesis
Lysosome	_____	h. cilia/flagella movement
Rough ER	_____	i. lipid synthesis/metabolism
Vacuole	_____	j. store/transport proteins

2. Eukaryotic vs. prokaryotic cells
    i. Differentiate their structures.

    ii. Give examples of other eukaryotic cells.

    iii. Give examples of other prokaryotic cells.

3. Explain the function of the following:
    i. Plasma membrane

    ii. Cytoplasm

4. List the differences between animal and plant cell structures.

6 LABORATORY

Animal Cells and Tissues

A. Instructional Objectives

After completing this exercise, the student should be able to identify body tissue types from slides and/or drawings. The student should also be able to name the body location of the tissues, know the general function of the tissues, and be able to give a definition of the following: all tissues introduced in the exercise, matrix, plasma, biconcave disc, whole mount, cross-section, cilia.

B. Introduction

The term "tissues" is defined as a group of cells, all with the same general function. It is a collection of similar cells, along with their intercellular material. Tissues are seen in multicellular organisms, with a division of labor among the cells. This leads to greater efficiency of overall function, because tissues are specialized to perform specific functions. There are 4 major classifications of tissues:

1. Epithelial—Tissues classified this way are composed primarily of cells, packed closely together, often forming a sheet of tissue. There is very little intercellular material. Epithelium has a covering or lining function, because of the way in which the cells fit together. A good example of epithelium is the outer layer of skin. Epithelial tissue is named by describing the pattern of cells seen on a slide. If one can see only one layer of cells, the word "simple" is used; if multiple layers of cells can be seen, the word "stratified" is used. Next, the shape of the individual cells is described. If the cell looks flat, or tile-like, or reptilian (like the pattern on the back of a snake), the term is "squamous." If the cell looks almost or perfectly round, the term is "cuboidal." If the cell looks oval, or similar to a cylinder, or elongated, the term is "columnar." The descriptive terms are used together to classify epithelial tissue.

Adapted from *Laboratory Exercises for an Introduction to Biological Principles* by Gil Desha. Reprinted by permission of the author.

For example, simple squamous epithelium, simple cuboidal epithelium, simple columnar epithelium, stratified squamous, stratified cuboidal, and stratified columnar epithelia are six types. There is an unusual classification for one tissue type, called pseudostratified ciliated columnar epithelium.

2. Connective—Tissues classified as connective contain cells and extracellular matrix. They typically do not form a sheet of tissue, and thus do not have a covering or lining function. The greatest variety of cells are seen in this tissue classification. Extracellular matrix is nonliving material that exists between cells. The prefix "extra" means "outside of" in this example, outside of cells. The matrix is composed of ground substance and embedded protein fibers. The nature of the ground substance gives a strong clue to the name of the connective tissue, and its probable function. For example, ground substance can vary widely, from a soft, gel-like material to a hard, calcified one. To explain, if the ground substance is soft and gel-like, the connective tissue is adipose, or fat. Its function is protection and insulation. If the ground substance is liquid, the connective tissue is the plasma portion of blood, and its function is transportation. If the ground substance is firm, the connective tissue can be tendons, ligaments, or cartilage. The function is a binding one, meaning binding parts of the body together for support. Last, if the connective tissue is hard and calcified, the type is bone. The function is support and protection. The skeleton forms an overall framework for the body.

3. Muscle—This classification of tissue focuses on three muscle types. They are skeletal, cardiac, and smooth. Skeletal muscle is described as being voluntary, because the movement (contraction and relaxation) is controlled by the brain. It is under an individual's conscious control. Skeletal muscle is also described as being striated, which means under the microscope it looks striped to the eye. Cardiac muscle is also striated, but under the microscope a structure called the intercalated disc can be seen. The purpose of the structure is to insure that electricity passes from one cardiac muscle cell to the other in an extremely rapid fashion, so that the heart beats as one unit. This muscle tissue is involuntary, because under normal circumstances the beating of the heart is not controlled consciously. Smooth muscle tissue is involuntary, and nonstriated. It is seen in the walls of organs, such as the uterus, the bladder, the digestive system, and in the walls of arteries. The current lab exercise will require students to observe the skeletal muscle slide and the smooth muscle slide for examples of striated and nonstriated muscle tissue.

4. Nervous—This classification of tissue focuses on two cell types, the neuron and the neuroglial cell. Neurons are the basic functional units of the nervous system, and are designed to transmit electrical impulses from one cell to another. They have an asymmetrical appearance in the microscope and often do not resemble anything. Sometimes they look pyramidal, sometimes like a boomerang, sometimes oval, but they will always have one major characteristic in common. They have multiple extensions of their plasma membranes, called dendrites, and one axon, a long extension. Neuroglial cells are very small and round, in comparison to neurons, and much more numerous. Their function is to support the neuron; they do not transmit electrical signals.

C. Materials

Each student should have the following:

Microscope

Prepared slides of

    Simple squamous epithelium (the lung slide)

    Simple cuboidal epithelium (the kidney slide)

    Stratified squamous epithelium (the skin slide)

    Pseudostratified ciliated columnar epithelium (the trachea slide)

    Connective tissue (hyaline cartilage or the trachea slide)

    Connective tissue (the adipose slide)

    Connective tissue (the bone slide)

    Connective tissue (the blood slide)

    Muscle tissue—skeletal (the tongue slide)

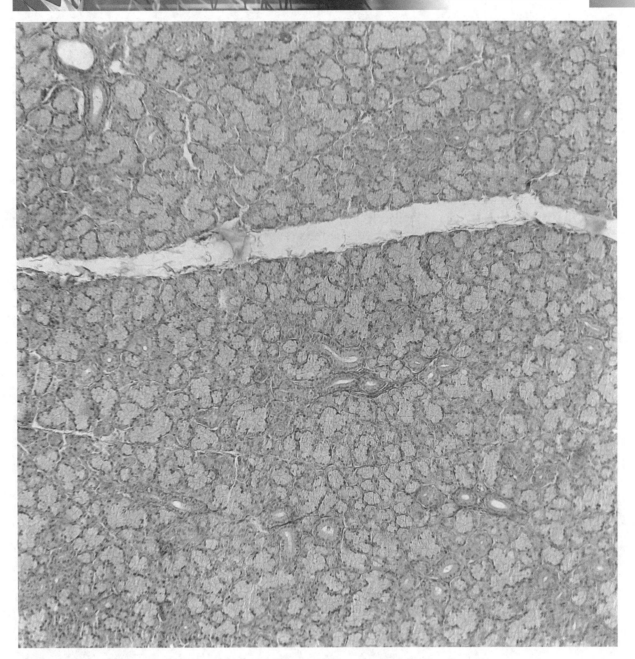

FIGURE 6.1   Simple Squamous Epithelium
© 2011 by Jubal Harshaw, used under license of Shutterstock, Inc.

Muscle tissue—smooth (bladder or frog duodenum slide

Nervous tissue (giant multipolar neuron smear slide)

D. Methods

Each student should examine the slides carefully. Only a small portion of many different types of animal tissues will be observed. A few of the slides will contain human cells, but those from other vertebrates are so similar that it takes a highly trained individual to make a distinction. Students will receive a closely related picture of the human body by observing comparable tissues of other animals.

1. Squamous epithelium—Students have two slides of squamous epithelium to observe, the lung slide and the skin slide. First, take the lung slide and put it in the microscope to see an example of simple squamous epithelium. Start with the scanning lens, and focus on the slide—the student might be reminded of looking at "old lace." Continue focusing with higher magnification until high magnification (100x)

is reached. Since the lung slide is examined, this type of epithelium lines cells called alveoli. They are the cells of gas exchange in the lungs, where oxygen and carbon dioxide are exchanged. Notice only one layer of cells is seen here, and the shape of the cells is flattened. The cell membranes are distinct against the lighter color of the cytoplasm. The darker stained nucleus can be seen inside the individual cells. This type of tissue is also found lining the human body cavity (coelom), covering the internal organs, and forming the mesenteries (around intestines) and Bowman's capsule (in the kidney). Next, take the skin slide to observe **stratified squamous epithelium**. This type of epithelial tissue is also located lining the esophagus, and lining the inside of the mouth. This tissue is clearly multilayered, as can be seen by looking at Figure 6.2. The outermost layer will be flat, or squamous; beneath these, the cells are more of a cube shape. These types of cells have what is known as "high mitotic potential," meaning they are constantly undergoing mitosis, dividing to renew and replenish themselves. Notice the darkly stained nuclei of these cells, but the individual cell membranes are not distinct. This tissue type will typically be seen at the top edge of the strip of skin, and all the cells in it will stain a darker color than the rest of the skin strip.

2. **Simple cuboidal epithelium**—Use the kidney slide to see this type of epithelium. At first, when focusing with the scanning lens, the student will see a mass of cells, with no distinct patterns. Focus with low power magnification, and the student will see cells arranged in circle forms, with each individual circle being formed by a group of round cells. The nuclei will be darkly stained and easy to see. Each individual cell is almost perfectly round, thus the term cuboidal is used. There is only one layer of these cells, thus the term *simple* to describe how many layers are seen. This cell type lines

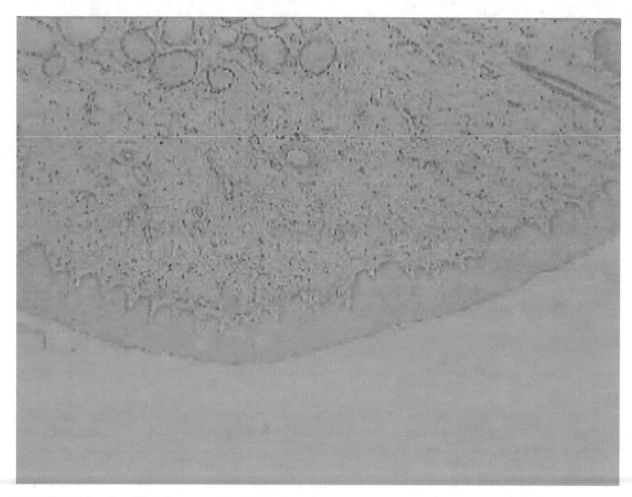

FIGURE 6.2   Stratified Squamous Epithelium
Copyright © William A. Olexik.

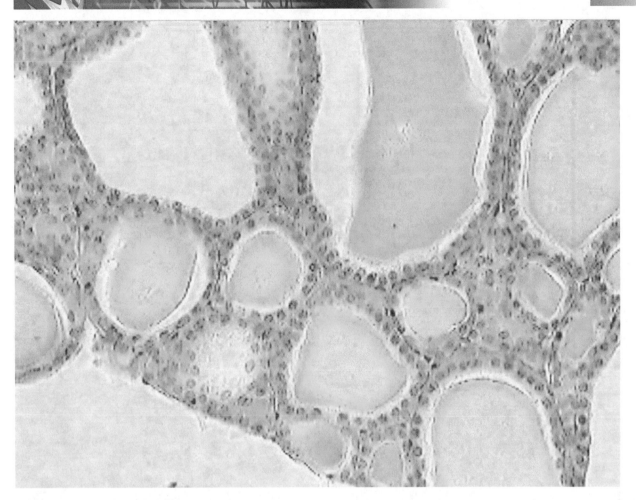

FIGURE 6.3   Simple Cuboidal Epithelium
Copyright © William A. Olexik.

the kidney ducts, but the student is looking at cross-sections, so the epithelial cells look as if they are forming a circle. The function is secretion of substances that will become urine. Observe Figure 6.3.

3. Ciliated columnar epithelium—Use the trachea slide to view this kind of epithelium. The full name is **ciliated pseudostratified columnar epithelium**, and this tissue lines the inside of the trachea. First, the term ciliated is used to describe tiny cilia at the very edge of the tissue. Their function is to move in a beating, back-and-forth motion, to trap and collect any dust or particles to keep them from going into the lungs. The cilia will appear as tiny hairs (like eyelashes) in the microscope. Look at Figure 6.4. Now look carefully at the cells. They are classified as columnar because individually they look elongated. They are all attached to a base, and are arranged in a very tight formation, close to one another. But the tops of these columnar cells are all of varying heights, and thus appear to be stratified when they are not. Thus, the term pseudostratified is used to describe simple columnar cells that look as if they are stratified.

4. Connective tissue—**hyaline cartilage**. Use the trachea slide to view this tissue. Also recognize that this tissue can be found in the fetal skeleton and the costal cartilages of the ribs. Look directly in the middle of the trachea slide, and focus with low power magnification. Notice the cells appear mostly in pairs, as if two cells were just preparing to divide. These are chondrocytes, which are the cartilage tissue cells that are stained in the trachea slide. They are at the end of mitosis and are getting ready to split, thus two are typically seen together. The nuclei are prominently stained. The function of this tissue is support, to keep the trachea (airway) open. Observe Figure 6.5.

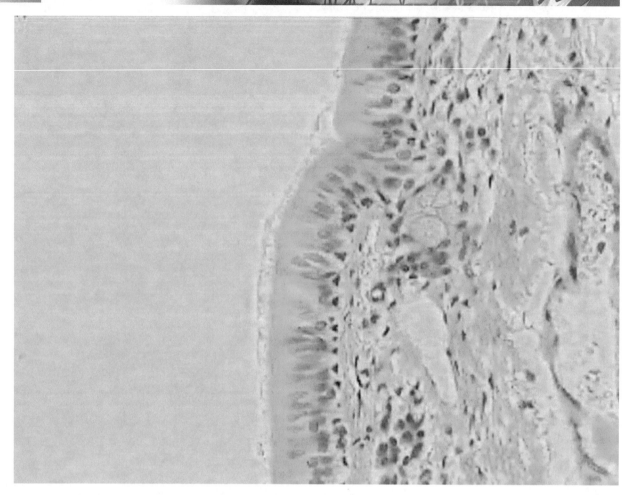

FIGURE 6.4   Pseudostratified Ciliated Columnar Epithelium
Copyright © William A. Olexik.

5. Connective tissue—**adipose**, or fat cells. This tissue is very easy to recognize, as long as it is stained well. It looks like chicken wire. Sometimes, the stain is weak and difficult to see. Use the iris diaphragm lever to decrease the amount of light flooding the slide, as the scanning lens is used to focus. The adipose tissue is found under the skin, around the heart and kidneys, and around the ovaries. The function is protection and insulation. Observe Figure 6.6.

6. Connective tissue—**bone**. Use the bone, dry ground slide. Typically the scanning lens is sufficient to see the cross-section of a long bone, which will resemble a tree trunk that has been cut. The compact bone tissue is made of a series of concentric rings, each one called a Haversian canal system. The purpose of the canals is to allow arteries, veins, and nerves to supply the living bone cells. The cells are surrounded by hardened calcium salts which are secreted by the cells. The bony matrix is composed of calcium salts. Observe Figure 6.7.

7. Connective tissue—**blood**. Use the human blood smear slide. Blood cells are composed of formed elements (about 45%), and the liquid portion called plasma. The formed elements are the three cell types—red blood cells, white blood cells, and platelets. The biological terms are erythrocytes, leukocytes, and thrombocytes, respectively. Observe Figure 6.8.

   a. Red blood cells—Red blood cells are biconcave discs, and contain a pigment called hemoglobin. Their function is to bind and carry oxygen through the arteries from the heart to body tissues. In a normal healthy adult, there will be 4.5 to 5.5 million per cubic mm of blood. In the slide, almost all of the cells will be of this type. They do not contain nuclei, and will live approximately 120 days. Thus, they must constantly be renewed by the body.

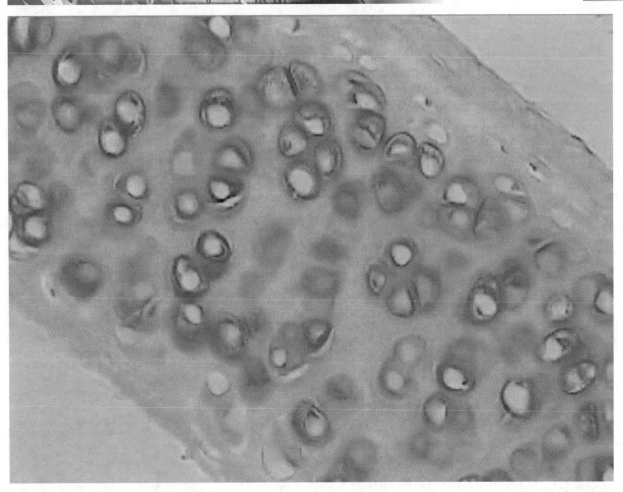

FIGURE 6.5 Hyaline Cartilage
Copyright © William A. Olexik.

    b. White blood cells—Typically, there will be 5,000 to 9,000 cells per cubic mm of blood. Their nuclei stain a dark purple color, and the cells are much larger than the red blood cells. They function in the immune system.

    c. Thrombocytes—These are actually bits and pieces of much larger cells. They are extremely small (2 to 3 microns) and are difficult to see, appearing as blue dustlike particles around the other cells on the slide. There are 250,000 to 450,000 per cubic mm of blood. Their function is blood clotting. Look at Figure 6.8.

8. Muscle tissue—

    a. First, use the tongue slide for striated, voluntary muscle tissue. Look at Figure 6.9 here. This tissue is composed of elongated cells called muscle fibers. During development, the individual muscle cells fuse together into units surrounded by a common membrane called the sarcolemma. Under high power (100x), small striations, or stripes, can be seen running at right angles to the long axis of the fibers. The fibers appear pink in color, and the nuclei are stained purple, found beneath the sarcolemma. Each fiber is made up of many smaller myofibrils running the length of the fiber. The function of skeletal muscle tissue is movement of the bones, caused by the contraction and relaxation of the muscles attached to them by tendons.

    b. Next, use the frog duodenum, or bladder slide to see smooth muscle. It is composed of individual cells which are elongated, and tapered towards a point at each end. The muscle cells in the duodenum slide (small intestine) will be running in a circular pattern, because the slide is a cross-section of the frog's small intestine. The individual smooth muscle cells will possess one darker stained, elongated nucleus. Look at Figure 6.10 here. The function of smooth muscle tissue is contraction and relaxation of the walls of organs.

9. Nervous tissue—Use the giant multipolar neuron smear to observe nerve and neuroglial cells. The slide is prepared from the spinal cord. The neurons are much larger than the neuroglial cells, and will stain blue, with a darker stained nucleus in the center of the cell body. The shape of the cell body will vary widely, but will have extensions of the plasma membrane stretched out for signal conduction. The function of the neuron is to generate and conduct electrical signals, to and from the central nervous system. The extensions receiving information from another cell are called dendrites; the one long extension conducting information away from the cell body to another cell is called the axon. Look at Figure 6.11.

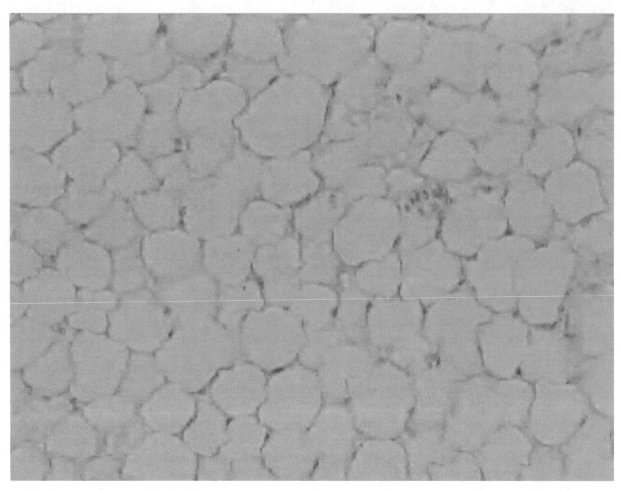

FIGURE 6.6   Adipose Connective Tissue
Copyright © William A. Olexik.

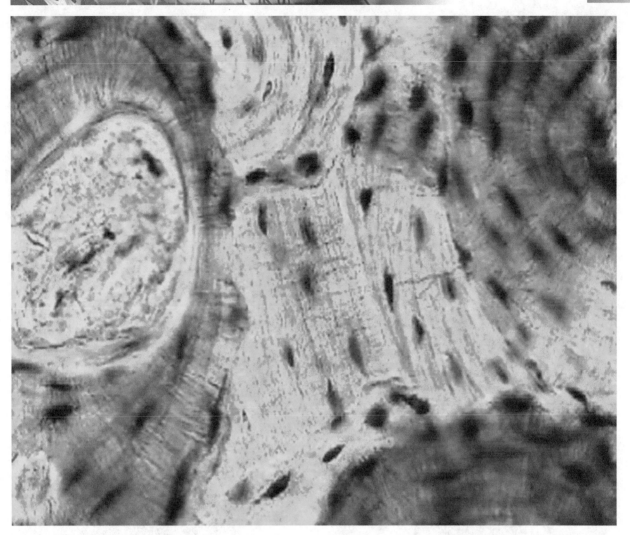

FIGURE 6.7   Bone
Copyright © William A. Olexik.

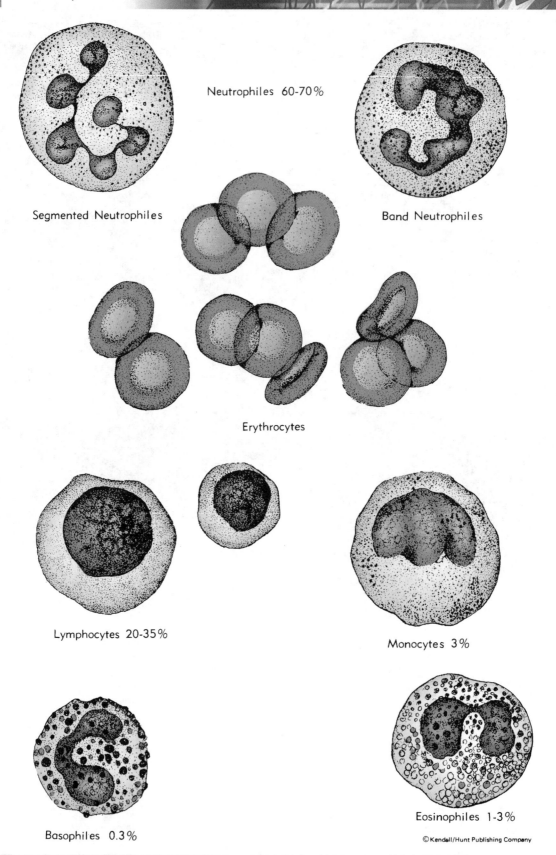

Neutrophiles 60-70%

Segmented Neutrophiles

Band Neutrophiles

Erythrocytes

Lymphocytes 20-35%

Monocytes 3%

Basophiles 0.3%

Eosinophiles 1-3%

FIGURE 6.8   Types of Blood Cells. (Percentages indicate relative frequencies of leukocytes.)

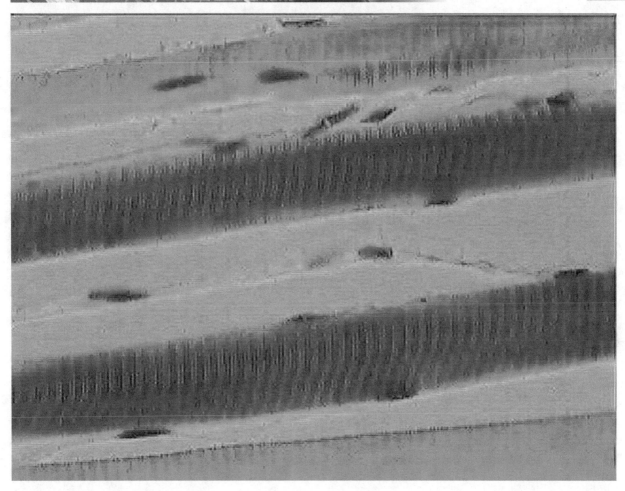

FIGURE 6.9 Skeletal Muscle
Copyright © William A. Olexik.

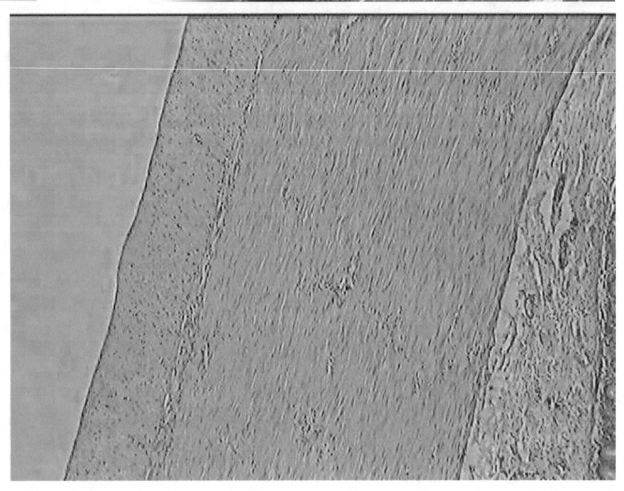

FIGURE 6.10   Frog Duodenum, Smooth Muscle
Copyright © William A. Olexik.

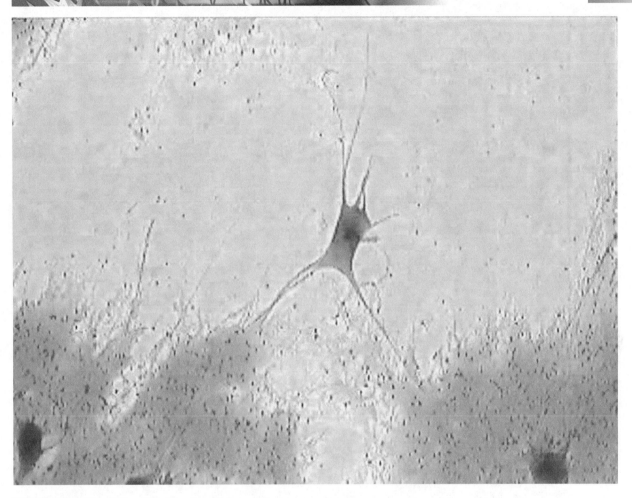

FIGURE 6.11 Giant Multipolar Neuron Smear

Name: _____

QUESTIONS

Complete the chart below, listing the tissues examined, their body locations, and general functions.

Tissue	Function	Location

7

LABORATORY

Organ Systems of the Body

A. Instructional Objectives: After completing this exercise the student should be able to:

1. Describe the anatomical position.
2. Know the definitions of and differentiate between organ and organ system.
3. Identify the following body cavities from a human torso model or dissected cat or rat, and be able to list organs occupying these cavities:
   i. Dorsal cavity
      1. cranial cavity
      2. spinal cavity
   ii. Ventral cavity
      1. thoracic cavity
         a. pleural cavity
         b. pericardial cavity
      2. abdominopelvic cavity
         a. abdominal cavity
         b. pelvic cavity
4. For any structure/organ covered in this exercise, know its function and which of the following organ system(s) it belongs to:

Integumentary	Digestive	Skeletal
Respiratory	Urinary	Nervous
Muscular	Reproductive	
Endocrine	Circulatory	

5. Identify on human torso model and discuss the following directional terms:
   i. **Dorsal**—the "backside" of a human or "upper surface" of an animal
   ii. **Ventral**—the "frontside" of a human or "belly surface" of an animal

    iii. **Anterior**—the "frontside" of a human or "head-end" of an animal

    iv. **Posterior**—the "backside" of a human or "tail-end" of an animal

    v. **Medial**—closer to the midline of the body

    vi. **Lateral**—further from the midline of the body

    vii. **Superior**—closer to the head in a human

    viii. **Inferior**—closer to the feet in a human

    ix. **Proximal**—closer to the point of attachment of appendage

    x. **Distal**—further from the point of attachment of appendage

6. Understand and be able to use the following body planes:

    i. **Coronal**—a longitudinal plane that divides ventral from dorsal

    ii. **Midsagittal**—a longitudinal plane that divides a structure into identical left and right halves

    iii. **Transverse or cross**—a plane that divides superior from inferior

7. Define, understand, and be able to use all terms in this exercise plus:

    i. Diaphragm

    ii. Mediastinum

    iii. Mammal

8. Complete the questions at the end of this exercise.

B. Introduction:

When working through this exercise, it is assumed that the body is in the **anatomical position**. In the anatomical position the person is standing erect with both feet shoulder width apart, facing the observer with both arms at the sides. The palms should be facing forward.

An **organ** is a collection of tissues and their secretions arranged in a specific way to produce a characteristic shape or structure. An **organ system** is a collection of organs serving a similar function. For example, the circulatory system is composed of the heart, vessels, blood, and lymph. All of these structures have their own function but work together to circulate blood to all parts of the body. Chart 7.1 lists the organ systems along with their organs/structures and their functions.

C. Materials:

Preserved cat and/or rat

Dissecting instruments

Human torso model

D. Methods:

Using the illustrations provided, the human torso models available, and the preserved specimen, locate all organs of the following organ systems.

CHART 7.1 Organ Systems

Organ System	Function(s)	Organs/Structures
Integumentary	Covers surface of body Protect against infection, water loss, sun rays, extreme changes in body temperature	Skin, hair, nails, horns, oil, sweat glands
Digestive	Breaks down food into chemicals that can be absorbed into the blood	Lips, tongue, teeth, salivary glands, esophagus, stomach, intestines, rectum, anus, liver, pancreas, gallbladder
Circulatory	Distributes blood to all portions of the body	Heart, blood vessels, lymph fluid
Respiratory	Ensures adequate oxygenation of blood Ensures proper removal of carbon dioxide from blood	Lungs, trachea, series of tubes that transport air into and out of lungs
Urinary	Formation of urine Remove waste products from the blood Homeostasis	Kidneys, ureters, urinary bladder
Skeletal	Protect organs Levers for movement Produce blood cells Site for calcium and phosphate storage and retrieval	Bones, cartilage, various skeletal membranes
Muscular	Movement	Skeletal, cardiac, smooth muscle
Reproductive	Produce offspring	Female-ovaries, uterus, fallopian tubes, vagina, clitoris, cervix Male-testes, scrotum, epididymis, vas deferens, urethra, prostate gland, seminal vesicles, bulbourethral glands, penis
Nervous	Communication	Brain, spinal cord, nerves
Endocrine	Relay messages between organs via hormones	Hypothalamus, pituitary, pineal gland, thyroid, parathyroid, thymus, adrenal, pancreas, testes, ovaries
Immune/ Lymphatic	Provide protection/defense from foreign invaders	Tonsils, bone marrow, lymph nodes, lymph vessels, spleen

Many organs are found in spaces called body cavities. The major cavities of mammals are illustrated in Figure 7.1.

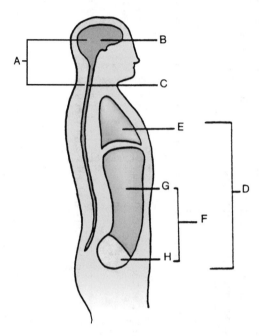

Place the correct letter of the indicated body cavity in the space provided.

_____ Abdominal   _____ Pelvic

_____ Abdominopelvic   _____ Spinal

_____ Cranial   _____ Thoracic

_____ Dorsal   _____ Ventral

FIGURE 7.1   Body Cavities

From *Anatomy and Physiology Laboratory Manual* 3rd edition revised printing by Robertson/Nabors/Lindsey/Barton, published by Kendall Hunt on page 3, figure 1.B.

Study Figure 7.2 using the previous descriptions of body planes and anatomical terms.

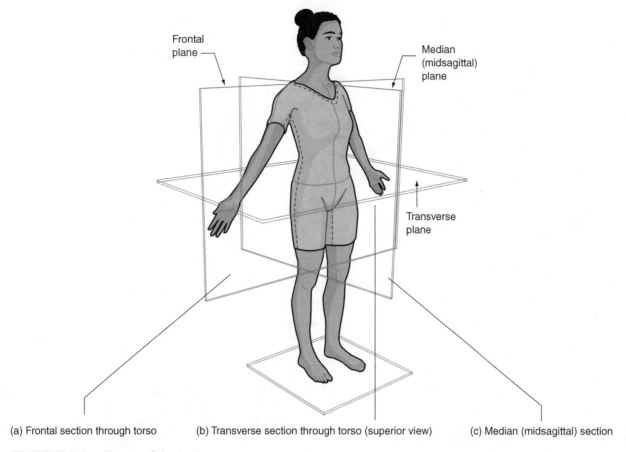

Frontal plane

Median (midsagittal) plane

Transverse plane

(a) Frontal section through torso   (b) Transverse section through torso (superior view)   (c) Median (midsagittal) section

FIGURE 7.2   Planes of the body

From *Human Anatomy and Physiology Laboratory Manual*. Cat Version. 8th ed. by Elaine N. Marieb. Copyright (c) 2005 by Pearson Education, Inc. Reprinted by permission.

Name: _____

QUESTIONS

1. Write the name of the organ system(s) to which each of the following organs/structures belongs:

   1. Liver _____

   2. Urinary bladder _____

   3. Kidney _____

   4. Lungs _____

   5. Hair _____

   6. Small intestine _____

   7. Seminal vesicle _____

   8. Heart _____

   9. Trachea _____

   10. Pancreas _____

   11. Spinal cord _____

   12. Ureter _____

   13. Adrenal gland _____

   14. Diaphragm _____

   15. Cartilage _____

   16. Urethra _____

   17. Testes/ovaries _____

   18. Pituitary _____

2. Write the name of the body cavity that each of the following organs/structures is found in.

   1. Brain _____

   2. Stomach _____

   3. Urinary bladder _____

   4. Lungs _____

   5. Heart _____

6. Liver _____

7. Spleen _____

8. Spinal cord _____

9. Fallopian tubes _____

10. Gallbladder _____

3. Fill in each blank with the best directional term.

1. The nose is _____ to the mouth.

2. The eyes are _____ to the nose.

3. The spinal cord is _____ to the shoulder blades.

4. The fingers are _____ to the elbow.

5. The back of the head is _____ to the face.

6. The palm of the hand is _____ to the back of the hand.

7. The hip is _____ to the toes.

8. When you bend over to touch your toes, you are bending along the

   _____ plane.

9. In order to divide the body into identical halves, you must draw a line down the

   _____ plane.

10. The _____ plane divides the anterior surface from the posterior

    surface of a man.

8 LABORATORY

The Cardiovascular System

I. Instructional Objectives

  a. Identify and explain the structure and function of the heart from models, drawings, and/or the sheep heart:

    i. External anatomy

      Pericardium

Coronary artery	Cardiac vein
Right ventricle	Left ventricle
Right auricle	Left auricle
Superior vena cava	Inferior vena cava
Pulmonary artery	Pulmonary vein
Aorta	

    ii. Internal anatomy

Right atrium	Left atrium
Right ventricle	Left ventricle
Tricuspid valve	Bicuspid valve
Papillary muscl	Chordae tendinae
Pulmonary semilunar valve	Aortic semilunar valve
Interventricular septum	Myocardium

  b. Explain functions of the heart by understanding:

    i. Method to measure pulse

    ii. Effect of exercise on pulse and blood pressure

    iii. Use of stethoscope

    iv. How heart sounds are produced and the events to which they relate

    v. How blood pressure is measured and how to read a sphygmomanometer

  c. Differentiate between an artery and vein in cross-section.

  d. Explain directional flow of blood in arteries and veins in relation to heart.

e. Identify the following systemic arteries and veins and locate them in the body.

Carotid artery   Jugular vein
Subclavian artery   Subclavian vein
Brachial artery   Brachial vein
Hepatic artery   Hepatic vein
Renal artery   Renal vein
Spermatic (ovarian) artery   Spermatic (ovarian) vein
Dorsal (abdominal) aorta   Inferior vena cava
Iliac artery   Iliac vein
Femoral artery   Femoral vein
                                   Saphenous vein
                                   Cephalic vein

f. Define the following terms:

"lub"	"dup"	pulse
stethoscope	blood pressure (BP)	sphygmomanometer
systole	diastole	diastolic/systolic pressure
target heart rate	atrioventricular valve (AV)	
double pump	oxygenated	deoxygenated

II. Introduction:

The heart is a rugged piece of living machinery which is vital for many animals. It pumps and thus forces blood through a system of living vessels by means of muscular contraction. Simpler animals such as fish have only two chambers in their hearts. Amphibians have three chambers. Birds and mammals have four chambers. Figure 8.1 illustrates the difference in blood flow between these animals. Figure 8.4 illustrates the flow of blood through the human heart as a **double pump**. This exercise will focus on the human heart.

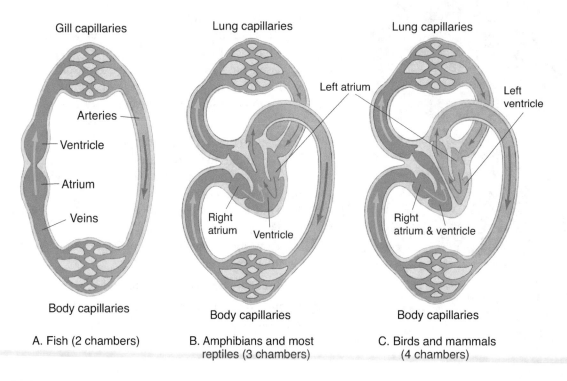

FIGURE 8.1   Vertebrate Circulation
© Kendall/Hunt Publishing Company.

The human heart begins its activity about the twentieth day of intrauterine existence and continues until death. It is a pressure pump that forces blood through a system of living vessels by means of muscular contraction. The contraction (pumping) phase of heartbeat is called systole. The relaxation phase is called diastole. The complete heartbeat, consisting of one systole and one diastole, takes about 0.8 seconds in the normal human adult. Mathematical calculation of this time into minutes shows 72 beats per minute to be the normal rate of heartbeat. The systolic increase in pressure can be felt in the arteries of the body as a pulse. The rate of the heartbeat is often referred to as the pulse rate. The pulse rate is regulated by both nerve and chemical stimuli and is affected by many factors in the internal and external environment.

Blood returns to and leaves the heart through a complex network of blood vessels. Veins return blood to the heart whereas arteries take blood away from the heart to the capillaries. Capillaries are microscopic blood vessels with walls that are permeable to the substances transported by the blood.

III. Materials:
   a. Microscope
   b. Dissecting tray
   c. Dissecting tools
   d. Mammalian heart
   e. Microscope slides of artery and vein
   f. Stethoscope
   g. Sphygmomanometer
   h. Heart model
   i. Circulatory system/heart chart/model

IV. Methods:

   A. Structure of the mammalian heart
      i. External anatomy
         1. Using Figure 8.2, study the surface of the heart and locate the structures indicated in the following description.
         2. If still attached to heart, note the pericardium, a tough membrane that surrounds the heart. The space between the heart and pericardium (pericardial cavity) is filled with fluid that reduces the friction generated by the heartbeat.
         3. Identify the coronary arteries and **cardiac veins** passing diagonally across the heart. These vessels may be obscured by white lines of fat. The arteries deliver blood to the walls of the heart, and the veins return blood from the walls of the heart to the right atrium. At times in the human heart, the coronary artery becomes blocked by a blood clot or foreign material. This condition is referred to as a coronary thrombosis and can result in death of some cardiac tissue. This can in turn result in death.
         4. The majority of the heart is composed of two posterior, cone-shaped, muscular, thick-walled ventricles. The right and left ventricles are separated by a thick wall called the interventricular septum.
         5. The other two chambers of the four-chambered heart are located at the anterior or broad end of the heart. These are the right and left atrium. They are covered by small, flaplike structures known as the right and left auricles. The auricles (or atria) have thinner walls than the ventricles and are usually of a darker color.
         6. Observe the stubs of the various blood vessels entering and leaving the heart. These will be studied in the following sections.
      ii. Internal anatomy
         1. The heart has been dissected in such a manner that each of the two sections contains a portion of the right and left halves of the heart.
         2. By observing the nature of the ventricles, determine for each of the two sections of the heart, the right and left sides. The left ventricle is larger and has much thicker walls than the right ventricle.
         3. With the heart in such a position that the right ventricle is on the left, locate the anterior (superior) vena cava and the posterior (inferior) vena cava. This can best be done by probing inside the

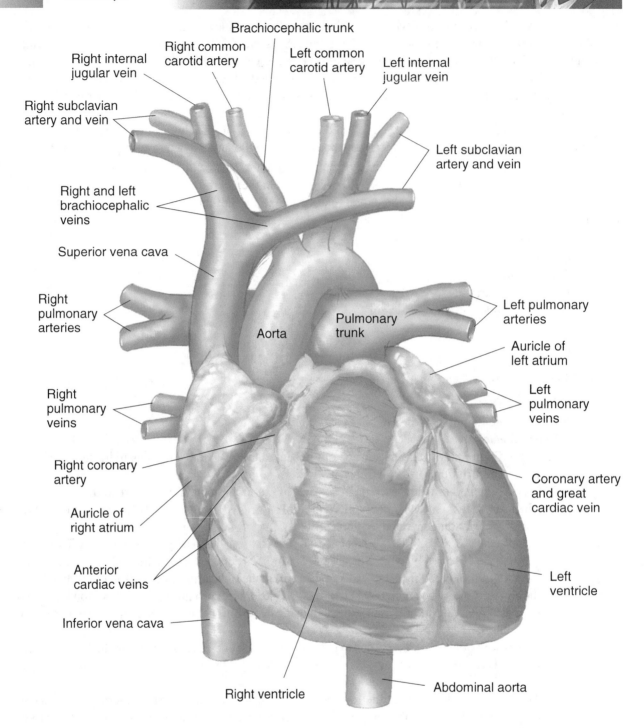

FIGURE 8.2   Anterior View of Heart
© Kendall/Hunt Publishing Company.

right atrium until two openings are located. Insert the probe through the openings and into the vena cavae. Notice that the anterior vena cava runs in a vertical position and that the posterior vena cava runs in a horizontal direction, almost 90° to the anterior vena cava. In the human, these vessels are usually called superior and inferior rather than anterior and posterior. The superior vena cava runs vertically and returns blood from the head and arm regions of the body. The inferior vena cava runs opposite to the superior vena cava and returns blood from other parts of the body.

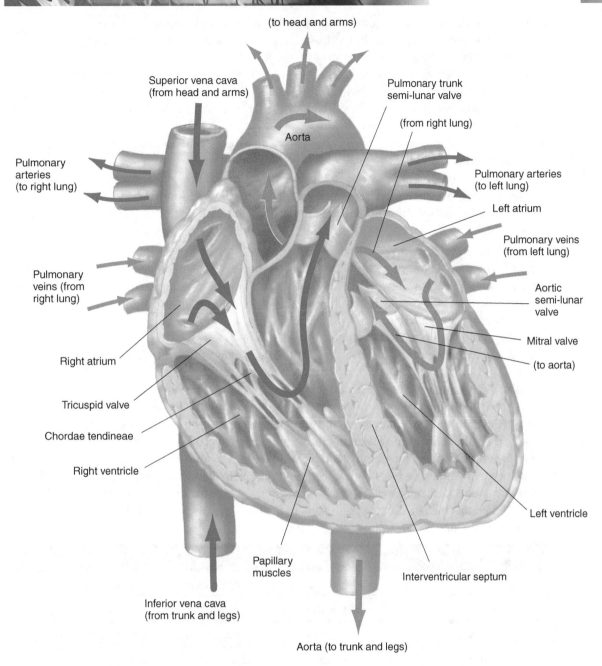

(to head and arms)

Superior vena cava
(from head and arms)

Pulmonary trunk
semi-lunar valve

(from right lung)

Aorta

Pulmonary
arteries
(to right lung)

Pulmonary arteries
(to left lung)

Left atrium

Pulmonary veins
(from left lung)

Pulmonary
veins (from
right lung)

Aortic
semi-lunar
valve

Right atrium

Mitral valve

(to aorta)

Tricuspid valve

Chordae tendineae

Right ventricle

Left ventricle

Papillary
muscles

Interventricular septum

Inferior vena cava
(from trunk and legs)

Aorta (to trunk and legs)

FIGURE 8.3　Interior Structures of Heart
© Kendall/Hunt Publishing Company.

4. Locate the three-flapped, membranous **tricuspid valve** situated between the right atrium and the right ventricle. This valve may be only partially intact due to the dissection and may appear as a thin, light colored fragment. This valve prevents the backflow of blood into the right atrium.

5. Observe the slender cords, **chordae tendinae** (heartstrings), which are attached to the tricuspid valve at one end and to the walls of the right ventricle at the other end.

6. Now locate the **pulmonary artery**. This can be done by probing in the right ventricle until an opening is found and by passing the probe through the opening. This vessel transports blood from the right ventricle to the lungs, where it branches into fine capillaries for the blood to receive oxygen.

7. Situated between the right ventricle and the pulmonary artery is the **pulmonary semilunar valve**. This valve is often obscured and difficult to find. It prevents the backflow of blood into the right ventricle.

8. Capillaries in the lungs become progressively larger and eventually develop into **pulmonary veins** that carry oxygenated blood from the lungs back to the heart. By using probes, locate the openings of the pulmonary veins into the left atrium.

9. Between the left atrium and left ventricle is a two-flapped bicuspid valve, also known as the **mitral valve**. As in the tricuspid valve, this valve may be only partially intact. Chordae tendinae are attached to the valve at the wall of the left ventricle. The bicuspid valve prevents the backflow of blood into the left atrium.

10. Using a probe, locate the **aorta**. This vessel distributes blood to arteries going to all body organs except the lungs.

11. Locate the **aortic semilunar valve** between the aorta and the left ventricle. This valve is usually severed, but observable. This prevents the backflow of blood into the left ventricle.

12. Compare the preserved mammalian heart with a model of the human heart.

B. Arteries and veins

i. Microscopic view

Observe comparable arteries and veins under low power of the microscope. Notice that arteries have thicker muscular walls and are less collapsed than veins. Arterial muscles allow for variation of diameter of these vessels based on need for blood. Draw an artery and a vein in the "Questions" section of this exercise.

ii. Pathway of blood through the heart and body

The human heart is a **double pump**. Half of the heart pumps blood to the lungs while the other half pumps blood to the other organ systems of the body.

Blood returning to the heart from the body is low in oxygen and is thus **deoxygenated**. The blood enters the right atrium of the heart from the superior and inferior vena cavae. It then passes through the tricuspid valve into the right ventricle. The right ventricle sends the blood to the lungs through the pulmonary trunk. The pulmonic semilunar valve prevents blood from "backing up" into the right ventricle. At the lungs, the blood picks up oxygen by way of diffusion from the pulmonary capillaries. Now that it is **oxygenated**, the blood returns to the left atrium of the heart through the pulmonary veins. It then passes through the bicuspid valve and into the left ventricle. The left ventricle then pumps the blood out the aorta and to the other organ systems of the body.

iii. Major systemic arteries and veins

Blood Vessel	Function
Aorta	Distributes blood to all systemic arteries
Superior vena cava	Returns blood to heart from veins of head/arms
Inferior vena cava	Returns blood to heart from veins of lower body and legs
Carotid arteries	Take blood to head
Jugular veins	Return blood from head
Subclavian arteries Subclavian veins	Take blood away from heart and towards arms Take blood from arm and towards heart
Cephalic veins (superficial)	Return blood from arms via subclavian veins
Brachial veins (deep)	Take blood to arms
Gastric artery	Takes blood to stomach
Hepatic artery	Takes blood to liver
Mesenteric arteries	Take blood to intestines
Hepatic vein	Returns blood from liver

Blood Vessel	Function
Renal arteries and veins	Take blood to and from kidneys
Spermatic (ovarian) arteries/veins	Take blood to and from gonads
Iliac arteries	Take blood to legs from trunk
Iliac veins	Take blood from leg towards trunk
Great Saphenous veins (superficial)	Return blood from legs via iliac veins
Femoral arteries	Major artery in the thigh
Femoral veins (deep)	Remove blood from legs

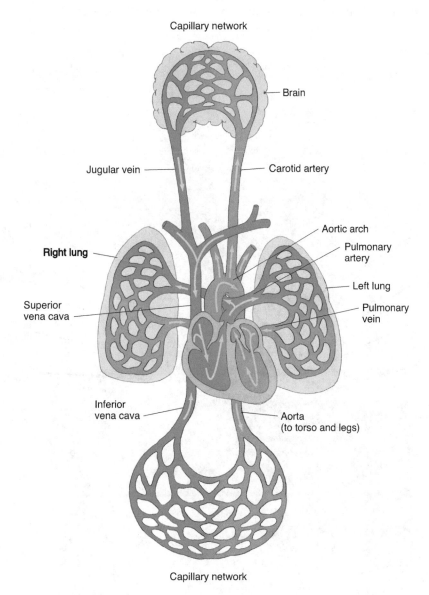

FIGURE 8.4   Diagrammatic Circulation

c. Functions of the heart

Students should work in pairs. One student should be the test subject and the other one should perform the test. Then the students should alternate.

i. Rate of heartbeat

When checking pulse, place the tips of the index and middle finger on the pulse site and press lightly. Determine the pulse rate per minute and record in the "Questions" section of the exercise. Check pulse at any of the following sites:

Radial—on wrist just lateral to the large tendon running down the center of the wrist.

Brachial—on the medial side of the upper arm in between the biceps and triceps.

Carotid—in the neck just lateral to the larynx. Be careful to check one side at a time, as checking both sides at once will deplete blood circulation to the head.

1. Test subject should be resting for at least five minutes; then measure the pulse rate while sitting at rest. Record in "Questions" section.

2. Exercise vigorously for two minutes and immediately check pulse rate. Then check pulse rate at two minutes and five minutes after exercise. Record these results.

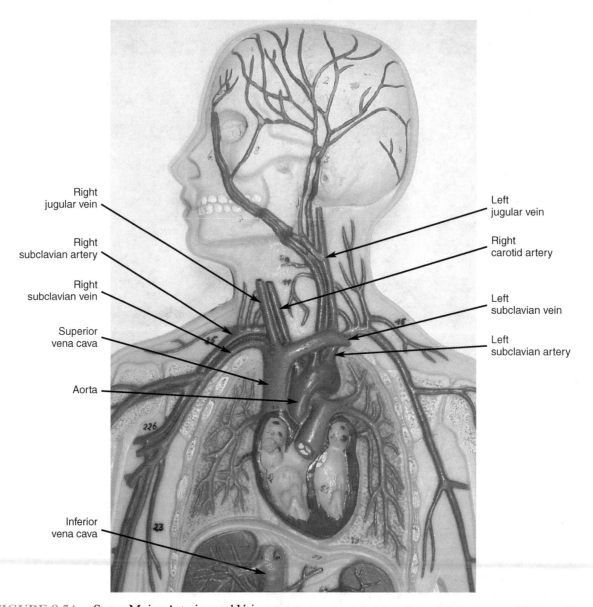

Right jugular vein

Right subclavian artery

Right subclavian vein

Superior vena cava

Aorta

Inferior vena cava

Left jugular vein

Right carotid artery

Left subclavian vein

Left subclavian artery

FIGURE 8.5A   Some Major Arteries and Veins. Reprinted by permission of The Science Source Company, Waldoboro, Maine.

ii. **Heart sounds**

Use a stethoscope to listen to the sounds of the heart. Two sounds should be distinguishable. The first sound is a low-pitched "lub" caused the by closure of the atrioventricular valves. The second sound is a short, sharp, high-pitched "dup." This results from the closure of the semilunar valves.

Feel the pulse while listening to the heart through the stethoscope.

iii. **Blood pressure (BP)**

1. The instructor will demonstrate the technique of using a sphygmomanometer
2. Test subject should be resting for at least five minutes; then check blood pressure.
3. Exercise vigorously for two minutes and immediately check blood pressure. Then check BP at two minutes and five minutes after exercise. Record these results.

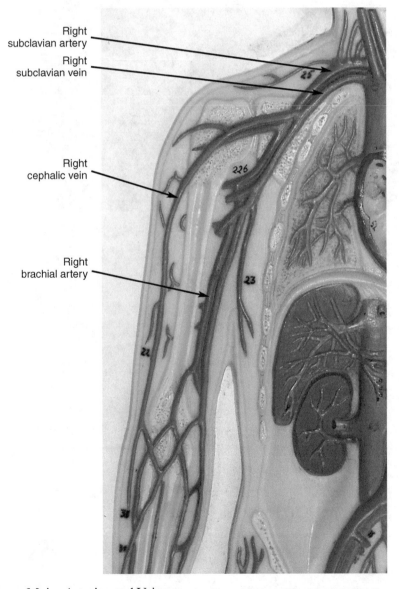

FIGURE 8.5B   Some Major Arteries and Veins. Reprinted by permission of The Science Source Company, Waldoboro, Maine.

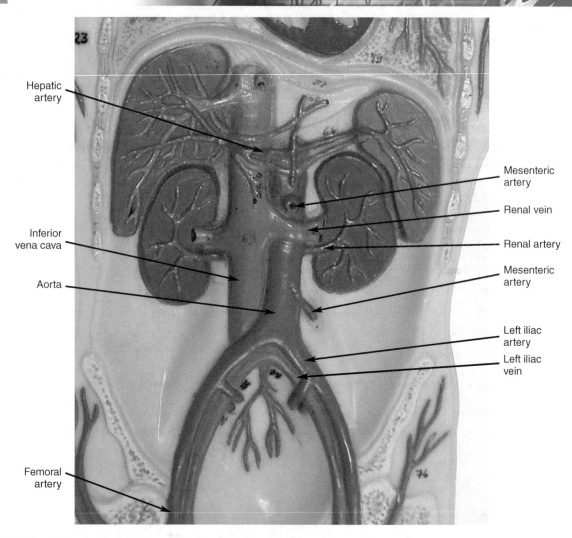

FIGURE 8.5C   Some Major Arteries and Veins. Reprinted by permission of The Science Source Company, Waldoboro, Maine.

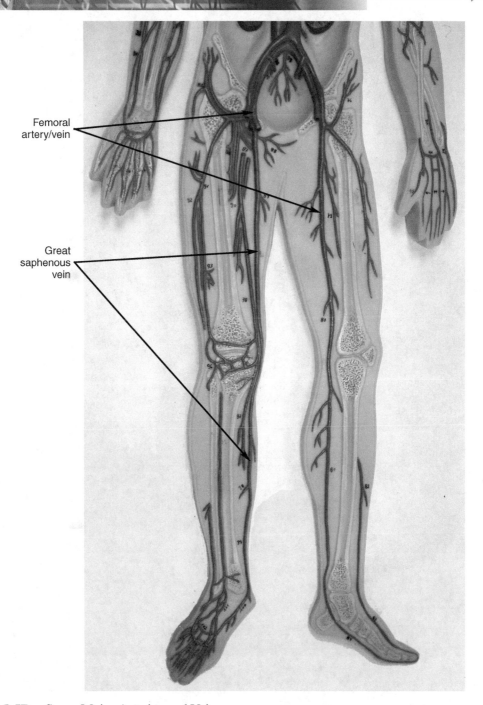

Femoral
artery/vein

Great
saphenous
vein

FIGURE 8.5D   Some Major Arteries and Veins. Reprinted by permission of The Science Source Company, Waldoboro, Maine.

Name: _____

E. QUESTIONS

a. Trace the pathway of blood through the heart. Indicate which structures carry oxygenated and deoxygenated blood.

    i. What kind of blood does the pulmonary artery carry?

    ii. What kind of blood does the pulmonary vein carry?

b. Explain why the left side of the heart is more muscular than the right side.

c. Distinguish between systolic pressure and diastolic pressure.

d. Draw an artery and vein (as seen in cross-section).

e. Effect of exercise on pulse rate and blood pressure:

Time	Pulse Rate	Blood Pressure
At rest		Systolic _____/_____ Diastolic
Immediately after exercise		Systolic _____/_____ Diastolic
Two minutes after exercise		Systolic _____/_____ Diastolic
Five minutes after exercise		Systolic _____/_____ Diastolic

f. What is your target pulse rate?

Age (years)	Average Maximum Heart Rate (bpm)	Target Zone: 60%-85% of Maximum
20	200	120 to 170
25	195	117 to 166
30	190	114 to 162
35	185	111 to 157
40	180	108 to 153
45	175	105 to 149
50	170	102 to 145
55	165	99 to 140
60	160	96 to 136
65	155	93 to 132
70	150	90 to 128

Reprinted with Permission. www.americanheart.org ©2010 American Heart Association, Inc.

How can knowing your target pulse rate be beneficial when exercising?

g. Name the veins that correspond to the following arteries.
   i. Renal artery   _____ vein.
   ii. Spermatic artery   _____ vein.
   iii. Pulmonary artery   _____ vein.
   iv. Gastric artery   _____ vein.
   v. Hepatic artery   _____ vein.
   vi. Carotid artery   _____ vein.

h. What is the function of capillaries?

i. What is the first organ to receive oxygenated blood pumped out of the left ventricle?

j. What is the function of the pericardium?

k. What is the function of the chordae tendinae?

9 LABORATORY

The Respiratory System

A. Instructional Objectives

After completing this exercise, the student should be able to

1. Identify and give the function of the following from models, drawings, or microscope slides.
   a. Nasal passage—nostrils, turbinate bones, nasal cavity, paranasal sinuses
   b. Pharynx and tonsils
   c. Larynx—epiglottis, glottis, and vocal cords
   d. Trachea and cartilage rings
   e. Bronchi, bronchioles, and alveoli
   f. Thoracic cavity, mediastinum, diaphragm, and lungs
2. Explain the mechanism of breathing based upon pressure changes occurring in the lungs, as movements of the diaphragm and ribs accompany inspiration and expiration. The student will use the bell jar apparatus and its parts.
3. Determine the percentage of normal vital capacity from information given in a specific example, such as height in inches of an individual, sex, age, the predicted vital capacity from a chart, and a spirometer reading. The student will know the diseases associated with a reduced vital capacity.
4. Explain the physiology of respiration. Relate the physiology to experiments conducted in the lab. The student will define and explain the following terms:
   a. Pons and medulla
   b. Hering-Breur reflex
   c. Aortic and carotid bodies
   d. Oxygen, carbon dioxide, and hydrogen ions, and reactions producing them
   e. Hyperventilation
5. Explain gas transport by oxyhemoglobin and bicarbonate ions, and explain gas exchange in the alveoli

Adopted from *Laboratory Exercises for an Introduction to Biological Perspectives* by Gil Desha. Reprinted by permission of the author.

6. Give definitions of the following:
    a. internal respiration
    b. external respiration
    c. tonsillitis
    d. laryngitis
    e. pleural cavity
    f. visceral and parietal pleura
    g. inhalation
    h. exhalation
    i. heme group
    j. spirometer

B. Introduction

The human respiratory system consists of the nose, nasal cavity, pharynx, larynx, trachea, bronchial tubes, and lungs. These structures function to allow atmospheric oxygen to enter the blood, and carbon dioxide from the blood to enter the atmosphere. The actual exchange of these gases occurs within the alveoli of the lungs. The exchange of the oxygen and the carbon dioxide between the blood and the tissues is called **internal respiration**.

C. Materials

Human torso, and other models
Bell jar/balloon apparatus
Spirometer
Paper bags and straws
Lime water
10 beakers (100 mL)

D. Methods

Part I—The Anatomy of the Respiratory System

Please refer to Figures 9.1, 9.2, and 9.3 as this section is read.

1. Nasal passage—Air enters the nostrils or nares where it passes into the nasal cavity. Once inside, the air is warmed, filtered, and humidified as it passes over the nasal membrane. The membrane covers the highly scrolled turbinate bones. Ciliated columnar epithelium forms the nasal membrane. This epithelium functions to sweep dust particles and bacteria into the throat, where they can be coughed out. The paranasal sinuses open into the nasal cavity. Sinuses are paired cavities located in the frontal, maxillary, sphenoid, and ethmoid bones of the skull. If they become infected, the condition is called sinusitis.

2. Pharynx—As air leaves the nasal cavity, it enters the back of the throat, or pharynx. This is a common passageway for air, liquid, and food. The tonsils, a type of lymphoid tissue, are located in the pharynx. If they become infected, this is a common condition called tonsillitis.

3. Larynx—Air passes from the pharynx to the larynx (or voicebox) by a slitlike opening called the glottis. The epiglottis is a leafy-shaped piece of tissue that covers the glottis as swallowing occurs. This functions to insure that liquid and solids go from the pharynx to the esophagus (and then to the stomach). The opening to the trachea is closed by the action of the epiglottis, and thus, remains clear when swallowing occurs. The internal surface of the larynx has delicate folds of mucous membranes which function as vocal cords. Altering the tension of these cords by laryngeal muscles as exhaled air passes over them (in conjunction with tongue activity), produces various sounds recognizable as words. Laryngitis is a condition caused by an inflamed larynx.

4. Trachea—This tube, about 4.5 inches long, passes from the larynx to the thoracic cavity, where it divides into two main branches, the primary bronchi. The trachea and the larger bronchial tubes are supported by C-shaped cartilaginous rings which prevent them from collapsing. Their function is to support the trachea to keep the airway open. Ciliated columnar epithelium lines the inside of the trachea, and sweeps dust and bacteria into the pharynx, where they can be coughed out.

5. Bronchi and alveoli—There are several successive branches from the trachea, called the bronchi. The initial two major branches are the primary bronchi. Branches from primary bronchi are called secondary

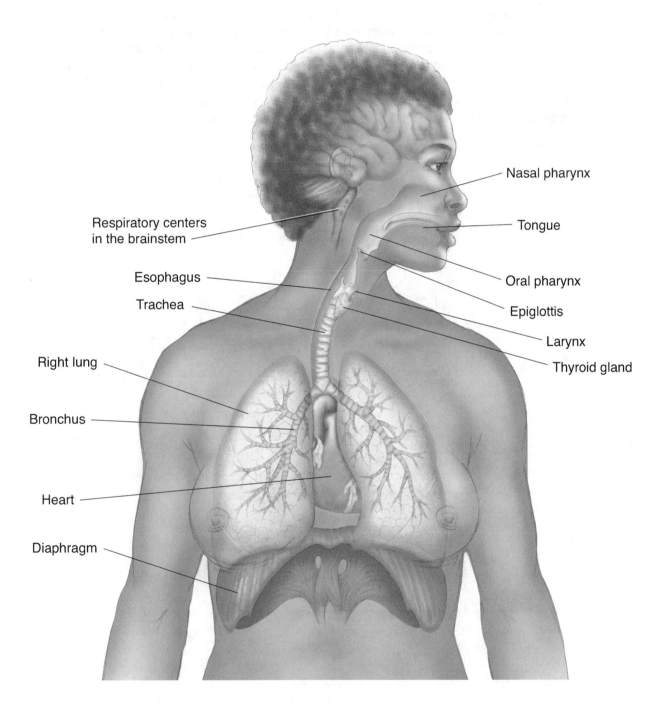

FIGURE 9.1   Respiratory System

Copyright © Kendall/Hunt Publishing Company.

bronchi. As branching continues to form numerous, branching tubes of smaller diameter, the bronchi lose their cartilaginous support and permeate all parts of the lung tissue as bronchioles (tiny bronchi). The smallest bronchioles terminate as alveolar ducts. Alveoli form the walls of these ducts as grapelike clusters. There are approximately 150 million alveoli, possessing about 350,000 square cm of surface area per lung. The alveolar walls are composed of simple squamous epithelium, which are surrounded by capillary beds. Gas exchange occurs across the alveolar wall, between the alveoli and lung capillaries.

6. Lungs—The lungs are contained in a body cavity called the thoracic cavity, which also contains the heart. The mediastinum divides the thorax into right and left thoracic cavities. The diaphragm, a membranous-muscular structure, separates the thoracic cavity from the abdominopelvic cavity, which lies inferior to the thoracic cavity. The diaphragm is the primary breathing muscle in the body. The lungs are specifically contained in the pleural cavities, which are lined with a delicate membrane called the parietal pleura. This same kind of membrane is also found on the outside of the lungs, and is called the visceral pleura when found there. The small cavity between the pleural membranes contains a small amount of pleural fluid, which functions to reduce friction as inhalation and exhalation occur. The membranes adhere to one another, thus, as the parietal pleura is expanded or contracted, the visceral pleura moves and brings about pressure changes that cause air to enter or leave the lungs.

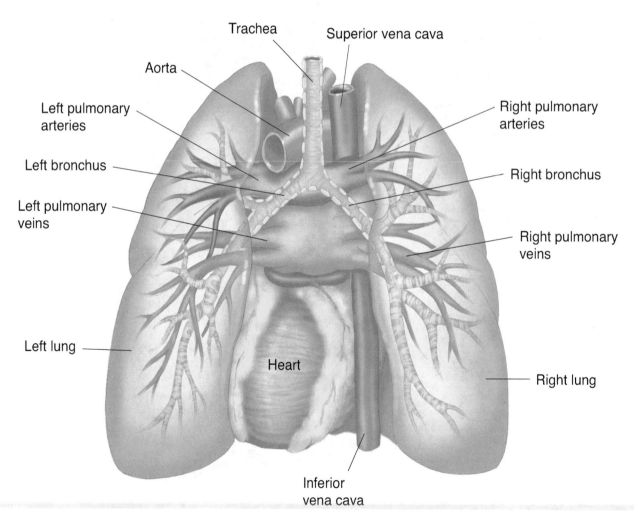

FIGURE 9.2   Posterior View of Lungs and Heart

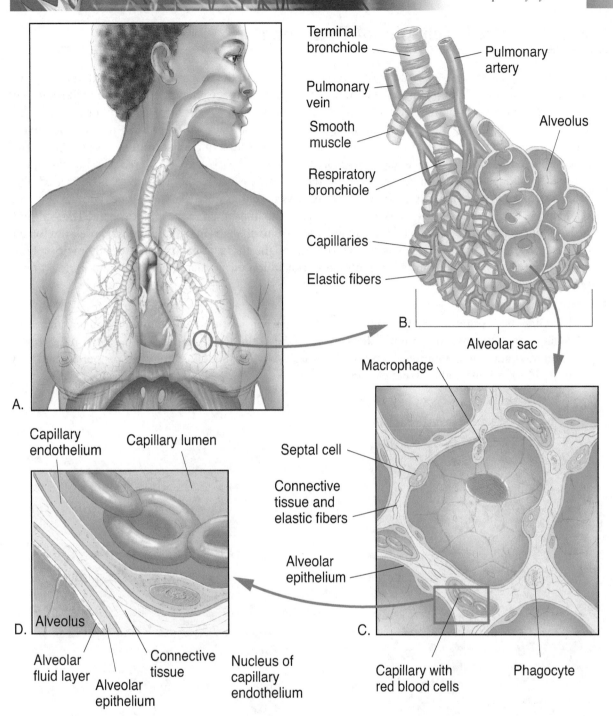

Terminal bronchiole

Pulmonary artery

Pulmonary vein

Smooth muscle

Respiratory bronchiole

Alveolus

Capillaries

Elastic fibers

B.

Alveolar sac

A.

Capillary endothelium

Capillary lumen

Macrophage

Septal cell

Connective tissue and elastic fibers

Alveolar epithelium

D. Alveolus

Alveolar fluid layer

Alveolar epithelium

Connective tissue

Nucleus of capillary endothelium

C.

Capillary with red blood cells

Phagocyte

FIGURE 9.3   Lung Anatomy

Part II—The Mechanics of Breathing

1. **Inhalation and Exhalation**—As the rib cage is elevated and the diaphragm is lowered, the lungs increase in volume to fill an increased amount of space in the thoracic cavity. The alveoli are elastic, and increase their internal volume. An air pressure gradient is established. There is more air pressure outside the lungs than inside the lungs. Air always follows its pressure gradient, flowing from an area of higher pressure towards an area of lower pressure, thus air enters the lungs from the atmosphere. This phase of respiration is called inhalation, or inspiration. When the rib cage is lowered, and the diaphragm elevated, the lungs decrease in volume because the volume inside the thoracic cavity is decreased. Part of the decrease is due to elastic recoil of the lungs. As the lungs decrease in volume, another pressure gradient is established where there is more pressure in the lungs than in the atomosphere. Thus, air will rush out of the lungs. Air enters and leaves the lungs, due to changes in the intrathoracic pressure, which is caused by increasing and decreasing the volume of the thoracic cavity. The intrathoracic volume changes are caused by the alternation contraction-relaxation of the diaphragm and rib muscles. Please refer to Figure 9.4.

2. **Bell Jar/Balloon Apparatus**
   a. Examine the respiratory model, consisting of a glass bell jar fitted with a rubber stopper, glass "Y" tubing, two balloons, and a rubber sheet fitted on the bottom of the model. Pull down on the rubber sheet, and allow it to return to its original position. Note results. Explain what happened, and why in terms of volume and pressure changes.

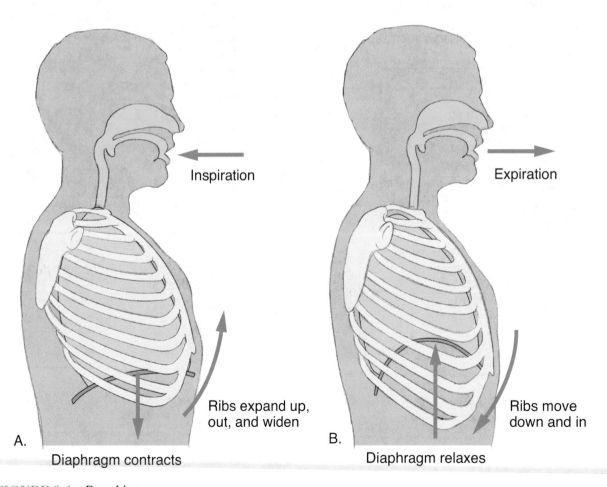

A.

Inspiration

Ribs expand up, out, and widen

Diaphragm contracts

B.

Expiration

Ribs move down and in

Diaphragm relaxes

FIGURE 9.4   Breathing

Copyright © Kendall/Hunt Publishing Company.

b. How does this compare with muscle action in the human body? What muscle movements bring about a similar action in the body?

c. Compare the bell jar model with the human system by naming the human counterpart that goes with the model part:

Model	Human Counterpart
glass tubing	?
balloons	?
rubber sheeting	?
walls of bell jar	?

3. **Vital Capacity Measurement**—An instrument called the spirometer measures vital capacity, defined as the amount of air forcibly exhaled from the lungs after forcible inspiration. Any lung disorder which does not allow the lungs to fill to maximum capacity, or to empty upon maximum exhalation, will reduce vital capacity. Emphysema and asthma are examples of two such conditions.

a. Set the spirometer dial to zero. Forcibly inhale, place the mouthpiece in your mouth, and close lips tightly. Now forcibly exhale, while holding nostrils closed. The spirometer has recorded the volume of forcibly exhaled air in mL. Repeat the process 3 times and record the highest spirometer reading as the actual vital capacity.

b. To determine the percentage of normal vital capacity for your sex, age, and height, consult the chart giving "predicted vital capacity." Then divide the actual vital capacity by the predicted vital capacity. Multiply that volume by 100 to get the percentage.

Actual vital capacity   _____ mL

Predicted vital capacity _____ mL

$$\% \text{ of normal vital capacity} = \frac{\text{Actual vital capacity}}{\text{Predicted vital capacity}} \times 100\%$$

c. Should you be concerned if your vital capacity is more than normal? Should you be concerned if it is less than normal? _____

d. How would emphysema affect vital capacity? _____

e. How would asthma affect vital capacity? _____

Part III. The Physiology of Respiration

1. **Regulation of Respiratory Movements**—The depth and rate of respiration are regulated by voluntary and involuntary mechanisms. Voluntary respiratory movements are under the control of the cerebral cortex. For example, one can hold his breath, or breathe deeply or shallowly, but it is impossible to have complete voluntary control over respiratory movements. Located in the pons and medulla of the brain are involuntary control centers that determine the rate and depth of respiration. The lungs, arterial levels of oxygen and carbon dioxide, and the pH of arterial blood are also part of the involuntary breathing mechanism.

As the lungs are filled with inspired air, stretch receptors within the lungs are stimulated. Impulses from these receptors are conveyed to the respiratory center in the medulla, and any further expansion of the lungs is inhibited. This mechanism is called the Hering-Breur reflex, which prevents overdistention of the lungs, and initiates the mechanism for expiration. The respiratory center of the brain initiates inspiratory movements, but when the lungs become stretched, the Hering-Breur reflex triggers expiration. If high arterial levels of carbon dioxide are detected by the aortic and carotid bodies, this stimulates increased respiration. Increased respiration is also brought about by a detection of very low levels of oxygen in arterial blood. This is usually accompanied by a decreased pH (acidic) of arterial blood. A decreased pH would mean more H+ (hydrogen ions) in blood volume, and this brings about increased respiration. High carbon dioxide levels are associated with increased blood acidity. Please note the chemical reaction below:

$$CO_2 + H_2O \rightarrow H_2CO_3, \text{ which dissociates in water to form } \rightarrow$$
$$\text{(carbonic acid)}$$

$$H+ + HCO_3^-$$
(hydrogen ion) and (bicarbonate ion)

2. **Variations in Respiratory Rate**—Have a lab partner count for you, and get the number of times per minute that you breathe for each of the situations listed below.
   a. _____ per minute, while sitting still
   b. _____ per minute, while running in place for 2 minutes
   c. Explain the basis for the change in respiratory rate.
   d. Breathe deeply and vigorously, with open mouth, for 1½ to 2 minutes. Is there a great urge to breathe immediately following this period of hyperventilation? Why?

3. **Voluntary Control of Breathing**
   a. How long can you hold your breath after a normal inspiration?
   b. After vigorous deep breathing?
   c. Summarize and explain your results.
   d. A simple test for carbon dioxide is to blow air through lime water. If a white material collects on the bottom of the beaker, carbon dioxide is present. Blow exhaled air into lime water $Ca (O_2H)$ in an aqueous solution.

What is the source of carbon dioxide? Does this confirm the presence of carbon dioxide in exhaled air?

4. **Respiratory Gas Transport and Exchange**
   a. Gas transport—Most of the oxygen in the blood is carried as a compound called oxyhemoglobin. Each RBC (red blood cell, or erythrocyte) normally contains about ¼ billion molecules of hemoglobin. Each hemoglobin (a red pigment) molecule contains 4 heme groups with an iron atom, which attracts and combines with one molecule of oxygen; theoretically, each RBC can carry a billion oxygen molecules.

      Most of the carbon dioxide in blood is carried as the bicarbonate ion, which is produced when carbon dioxide is combined with water in the blood. This forms the acid, carbonic acid, which dissociates in water to form hydrogen ions, and the bicarbonate ion. See previous figure if you cannot remember the chemical equation. The bicarbonate ions diffuse out of the red blood cells and into the blood plasma, where they are transported throughout the circulation.
   b. Gas exchange—In the body, carbon dioxide is picked up by venous blood at the level of the tissues. Carbon dioxide is released by tissues as a result of normal breakdown or oxidation of organic substances (food) in the cells. The tissues absorb oxygen out of the arterial blood, because oxygen is required for the cellular breakdown of food in a process called cellular respiration. Energy is released from this process, and is used to drive the work of the cell. So body cells are constantly consuming oxygen, and producing carbon dioxide. In the lungs, carbon dioxide diffuses from the blood into the alveoli, and at the same time, oxygen diffuses from the alveoli into the blood.
   c. Interpretation of gas chromatograph experiments—The chart on the following page summarizes chromatograph data. Study the data and answer the questions which follow.

Name: _____

QUESTIONS

Note: Amounts are in parts per thousand or cubic centimeters per liter.

	Oxygen	Carbon Dioxide
Amt. present in inhaled air	210.0	0.3
Amt. present in exhaled air	180.0	26.0
Amt. present in arterial blood	195.0	480.0
Amt. present in venous blood	150.0	520.0
Carrying capacity of blood plasma	0.3	
Carrying capacity of whole blood	195.0	

Use the information in the above chart, and circle the correct answer in each of the following statements.
1. There is more (a. oxygen; b. carbon dioxide) in inhaled air.
2. The lungs (a. do; b. do not) absorb all of the oxygen present in the inhaled air.
3. The oxygen content of (a. arterial; b. venous) blood is greater.
4. The lungs (a. do; b. do not) remove all the carbon dioxide from the blood.
5. The body (a. consumes; b produces) oxygen, and (a. consumes; b. produces) carbon dioxide.
6. Most of the oxygen is carried by the (a. blood plasma; b. red blood cells).
7. The lungs (a. absorb; b. release) oxygen, and (a. absorb; b. release) carbon dioxide.

10

The Digestive System

Part A

A. Instructional Objectives: After completing this exercise the student should be able to:

1. Identify on the human torso, the digestive system models, and a cat specimen and state the function of the following:

Oral cavity	Salivary glands*	Pharynx
Esophagus	Stomach	Small intestine**
Pancreas	Liver	Gallbladder
Colon	Appendix	Cecum
Rectum	Anus	

   *Parotid, submandibular, and sublingual

   **Duodenum, jejunum, ileum

2. Identify on the model of the intestinal wall and state the function of the following:

Villi	Blood capillaries	Lacteals

B. Materials

Models: human torso

intestinal wall

digestive system

C. Introduction

Humans are required to eat food to acquire nutrients to sustain them. Nutrients are substances that are essential to metabolism, the regulation of metabolism, maintenance of proper homeostasis, and growth and repair of the body. Proteins, lipids (fats and oils), carbohydrates, vitamins, minerals, and water are nutrients required by the body. Although some consider alcohol an essential part of their diet, alcohol is not a nutrient. It is a mind-altering drug!

For the human body to acquire the essential nutrients, the body must first ingest the food, digest food, and then absorb food. Digestion is the physical and chemical breakdown of food. Actually, mastication is part of the digestion process. Even with mastication, the nutrient molecules are too large to be absorbed by the wall of the digestive system. Enzymes are required to chemically break down the large nutrient molecules into smaller and more easily absorbed molecules. The nutrients then must be absorbed into the body through the circulatory system and the lymphatic system.

The body system that is involved in digestion and absorption is called the alimentary tract, also called the gastrointestinal (GI) tract. The alimentary tract or digestive system is a continuous tube that begins with the mouth and terminates at the anus. The digestive system has number of functions:

1. Ingestion
2. Mechanical digestion
3. Chemical digestion
4. Absorption of nutrients
5. Elimination of undigested or indigestible wastes

   Ingestion is simply eating and drinking. Ingestion means eating. Normally we eat food when we are hungry. Sometimes people eat for other reasons. The food that we like to eat has tastes and aromas that we find pleasurable.

   Mechanical digestion occurs throughout the digestive tract. It involves chewing in the mouth. Chewing involves the biting, shredding, and grinding of food so that the surface area is increased so that digestive enzymes can come in contact with nutrients. The stomach churns the food into a soupy mixture called chyme. The chyme enters the small intestine, which is segmented so that more digestion may occur.

   Chemical digestion involves enzymes. Enzymes are proteins that act as catalysts. Catalysts are substances that speed up a chemical reaction by bringing molecules into close contact. Digestive enzymes break the bonds of large nutrient molecules into smaller ones so that the body may absorb them.

   Absorption of nutrients primarily occurs in the small intestine, more specifically the duodenum and the jejunum. Water-soluble nutrients are absorbed into the blood through the blood capillaries. Lipids and fat-soluble vitamins are absorbed into lacteals, special lymph capillaries.

   Elimination is the removal of undigested or indigestible food (feces) through the anus.

D. Procedure
   1. Examine the human torso and locate the following structures; using your text, complete Table 10.1, Functions of the Digestive Organs.
      a. Mouth
      b. Salivary glands
      c. Esophagus
      d. Stomach
         (1) Pyloric valve
      e. Small intestine
         (1) Duodenum   (2) Jejunum   (3) Ileum
      f. Colon (large intestine)
         (1) Ileocecal valve   (2) Appendix   (3) Cecum
         (4) Rectum   (5) Anus
      g. Pancreas
      h. Liver
      i. Gall bladder
   2. Be able to distinguish between ingestion, digestion, and absorption.

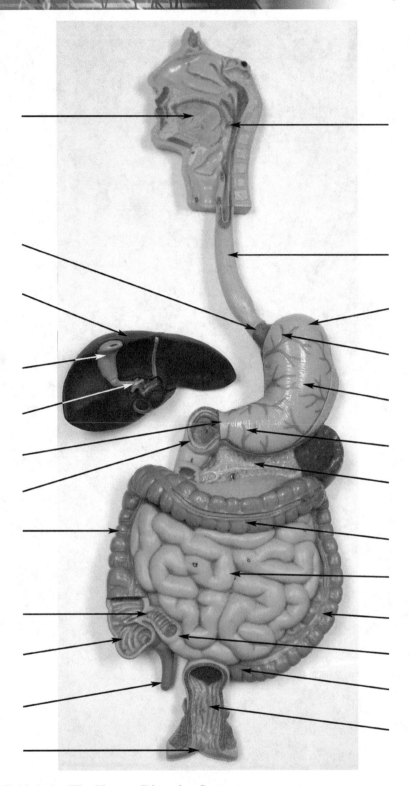

LABEL FIGURE 10.A.1   The Human Digestive System.

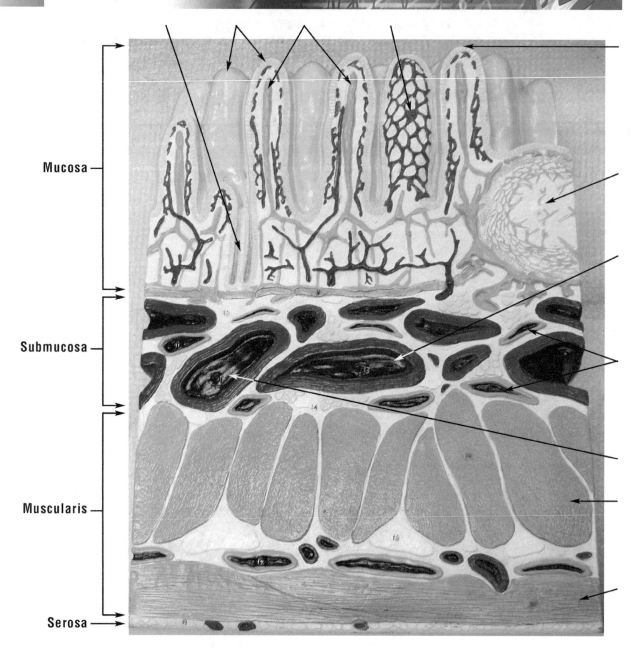

LABEL FIGURE 10.2.   From Anatomy and Physiology Lab Manual by Carolyn Robertson, Jerry Barton, II, Jerri Lindsey and Robert E. Nabors. Copyright © 2009 by Kendall Hunt Publishing Company. Reprinted by permission.

Part B. Digestion and Absorption of Food

A. Instructional Objectives
   1. Define the role of an enzyme in digestion.
   2. Distinguish between ingestion, digestion, and absorption.
   3. Distinguish between a substrate and an enzyme.
   4. Describe the action of the three classes of enzymes (carbohydrases, lipases, and proteinases).

TABLE 10.1 Functions of the Digestive Organs

Organ	Function
Mouth	
Teeth	
Tongue	
Salivary glands	
Esophagus	
Stomach	
Pyloric valve	
Small intestine	
Duodenum	
Colon	
Ileocecal valve	
Appendix	
Cecum	
Rectum	
Anus	
Pancreas	
Liver	
Gallbladder	

5. Identify the site of digestion, the gland or organ that secretes the enzyme, and the products of digestion for the following carbohydrases:
   a. Salivary amylase
   b. Pancreatic amylase
   c. Sucrase
   d. Lactase
   e. Maltase

6. Identify the site of digestion, the gland or organ that secretes the enzyme, and the products of digestion for the following proteinases:
   a. Rennin
   b. Pepsin
   c. Trypsin
   d. Chymotrypsin
   e. Carboxypepidase

7. Identify the site of digestion, the gland or organ that secretes the enzyme, and the products of digestion for lipase.

8. Explain the effects of temperature and pH on the activity of enzymes.

9. Explain the role of litmus and Benedict's solution in the experiment.

10. Explain the role of bile in fat digestion.

11. Explain the role of control in an experiment.

B. Materials

6 test tubes per lab team	400-mL beaker
Distilled water	Benedict's solution
5% pancreatic solution*	Blue litmus solution
1% starch solution**	Lugol's iodine solution (I_2KI)
Boiled eggs	Half-and-half
5% dextrose (glucose)	Red China marking pencils
Hot plate	Test tube clamps
Water bath (set at 40°C)	Heavy-duty Ziploc freezer bags

* Dissolve 5 g of powdered pancreatin in 100 mL of 0.5% $NaHCO_3$ solution.

** An easy way to prepare this solution is to spray a cheap brand of spray.

C. Introduction

The nutrients that humans eat are in the form of food groups known proteins, lipids, and carbohydrates. All of these nutrients contain the majority of molecules that are too large to be directly taken into the cells of our bodies. Therefore, these nutrients must be mechanically and chemically broken down into small molecules so that they can be absorbed and transported by our circulatory and lymphatic systems.

Chemical digestion is accomplished by the aid of enzymes. Enzymes are proteins that act as catalysts in chemical reactions. Remember that catalysts are substances that speed up a chemical reaction. Catalysts make the chemical reaction easier to occur by bringing the reactants closer together and causing certain chemical bonds to be broken and making others easier to form. In digestion the digestive enzymes interact with the substrate molecules and water to break them down into simpler molecules. The chemical process is called hydrolysis. In the process of digestion, enzymes break down these large molecules in the presence of water. There are three major classes of digestive enzymes:

1. Protienases break down proteins into amino acids.
2. Carbohydrases break down carbohydrates into simple sugars.
3. Lipases break down lipids into fatty acids and glycerol.

For instance, through enzymes, proteins are broken down into polypeptides and polypeptides are broken down into amino acids. Without enzymes, the body would not function properly; illness or death would be the next step. The following are some very important aspects of enzymes:

a. All enzymes are proteins; proteins are under the control of genes. Some enzymes are not functional due to mistakes in the genes. Lactose intolerance is due to faulty enzyme (lactase) that is responsible for breaking down lactose into glucose and galactose.
b. Enzymes are very specific; that is, they act only on a particular type of chemical bond in a specific type of molecule.
c. The specific substance upon which the enzyme acts is called the substrate. Splenda is made of sucralose, a sugar molecule that cannot be broken down by sucrase. The reason is that sucralose, though made of sugar, has a different shape than that of sucrose (table sugar).
d. Each enzyme has a specific set of optimal environmental conditions under which its activity is most efficient—that is, temperature, pH, and so on. For instance, pepsin is active only in the stomach, where the environment is strongly acidic, (pH $\approx$ 2). When the chyme is released into the duodenum, the environment is mildly acidic (pH $\approx$ 6.5); pepsin is inhibited in the duodenum. The pH of the jejunum and ileum is $\approx$ pH 7.5.
e. Tables 10.2 to 10.4 show the major enzyme classes.

The upper part of the small intestine absorbs the simple sugars (glucose, fructose, and galactose). The simple sugars are absorbed by the epithelial cells covering the villi. The absorbed simple sugars are absorbed by the blood capillaries in the villi and are transported via the hepatic portal vein to the liver. At the liver, some of the simple sugars are converted into glucose for energy production. Glucose is converted into

TABLE 10.2 Carbohydrate Digestion

Organ	Enzyme Secreted	Site of Enzymatic Activity	Substrate	Products
Salivary glands	Salivary amylase	Mouth	Starch	Maltose
Pancreas	Pancreatic amylase	Duodenum (small intestine)	Starch	Maltose
Small intestine	Maltase	Duodenum (small intestine)	Maltose	Glucose only
	Sucrose	Duodenum (small intestine)	Sucrose	Glucose + fructose
	Lactase	Duodenum (small intestine)	Lactose	Glucose + galactose

animal starch, glycogen. The human liver carries an 18-hour energy reserve in the form of glycogen. The excess simple sugars are converted by the liver into fat; fat then is stored in fat deposit areas of the body. Fructose has been found to stimulate the storage of fat. Fructose is found in foods and beverages that use corn syrup as a sweetener.

TABLE 10.3 Protein Digestion

Organ	Enzyme Secreted	Site of Enzymatic Activity	Substrate	Products
Stomach	Pepsin	Stomach	Proteins	Larger polypeptides
	Rennin	Stomach	Coagulated milk	Smaller polypeptides
Pancreas	Trypsin	Duodenum (small intestine)	Larger polypeptides	Smaller polypeptides
	Chymotrypsin	Duodenum (small intestine)	Larger polypeptides	Smaller polypeptides
Small intestine	Peptidases	Duodenum (small intestine)	Peptides	Amino acids

The lower part of the small intestine absorbs amino acids. The amino acids are absorbed into the blood capillaries and are carried via the hepatic portal vein to the liver and to all parts of the body for protein synthesis or for energy production. The liver may covert amino acids into other amino acids needed by the body. There are certain amino acids that cannot be made by the liver. These amino acids are called essential amino acids. The only way the body acquires these is through an adequate diet.

TABLE 10.4 Lipid Digestion

Organ	Enzyme Secreted	Site of Enzymatic Activity	Substrate	Products
Pancreas	Lipase	Duodenum	Fats	Fatty acids + glycerol

The liver does not produce any digestive enzymes. However, the liver produces bile. Bile is not an enzyme, but it is an emulsifying agent. Bile breaks fats into small microscopic droplets. This makes fat digestion easier by making the fat droplets so small that lipases can act upon the fat molecules. The lower part of the digestive system absorbs fatty acids and glycerol. Fatty acids are absorbed by the lacteals in the villi. The lymphatic veins carry the fatty acids to the liver. The fatty acids are used as energy source or stored in lipocytes (fat cells)

D. Procedure

To simplify the measurements, here is the rule of thumb for volumes:

1. The length of one thumb is about 5 mL.
2. The length of the joint of the thumb to the tip of the thumb is about 3 mL.
3. A thumb nail is about 1 mL.
4. All solutions should be at room temperature.

Part 1. Protein digestion

This is experiment is a demonstration. A hard-boiled egg, shelled, is placed in a heavy-duty Ziploc freezer bag and is allowed to stand immersed in a 5% pancreatin solution at room temperature for 24 hours. Another hard-boiled egg, shelled, is placed in a heavy-duty Ziploc freezer bag and is allowed to stand immersed in distilled water at room temperature for 24 hours. Record your observations in Data Table 10.1.

DATA TABLE 10.1   Protein Digestion

Medium egg is immersed in	Observations
Water	
5% pancreatin solution	

Part 2. Starch Digestion

1. Thoroughly stir the 1% starch solution before mixing.
2. Add the following solutions to the test tubes:

Test Tube	Carbohydrate Solution	1% Pancreatin Solution	Distilled Wate
A	5 mL 1% starch	3 mL	0 mL
B	5 mL 1% starch	0 mL	3 mL
C	5% glucose	0 mL	3 mL
D	0 mL	3 mL	5 mL

3. Place the test tubes in a warm water bath (40°C) for one hour.
4. Withdraw the test tubes and add 5 mL of Benedict's solution to each test tube. Place all the test tubes in a boiling water bath for about two minutes. Record your results in Data Table 10.2. Remember that Benedict's solution is a test for glucose. Glucose is formed when starch is chemically digested.

DATA TABLE 10.2   Observations of Carbohydrate Digestion

Test Tube	Solutions	Observations
A	1% starch solution + 5% pancreatin solution	
B	1% starch solution + water	
C	Glucose + water	
D	Water	

5. To test tube E, add a couple of drops of iodine (I_2/KI) to a freshly prepared starch solution. This is the test for the presence of starch. Record your results in Data Table 10.3

DATA TABLE 10.3   Observations for the Presence of Starch

Test Tube	Solution	Observations
E	1% starch solution	

Part 3. Digestion of Fat

Half-and-half is half fat butter fat (cream) and half milk. Litmus is an indicator for the presence of acids and bases. Blue litmus turns red in the presence of an acid. Red litmus turns blue in the presence of a base.

1. Add the following solutions to the test tubes:

Test Tube	Half-and-half	1 % Pancreatin Solution	Distilled Water	Blue Litmus
F	5 mL	2 mL	0 mL	Enough to make a distinct blue
G	5 mL	0 mL	2 mL	Enough to make a distinct blue of equal intensity

2. Place both tubes in the warm water bath (40°C) for about one hour.

3. Remove the tubes from the water bath and record your observations in Data Table 10.4.

DATA TABLE 10.4   Digestion of Fat Test Tube Observations

Test Tube	Solution	Observations
F	Half-and-half + 1% pancreatin	
G	Half-and-half + water	

Name: _____

QUESTIONS

Part 1. Protein Digestion

1. Describe the surface of the egg immersed in distilled water.

2. Describe the surface of the egg immersed in 1% pancreatin solution.

3. What evidence is there that the experimental egg is being digested?

4. What is the purpose of the control?

5. Name the pancreatic enzymes responsible for protein digestion.

6. Which organs produce enzymes that digest protein?

7. In which two organs does protein digestion take place?

Part 2. Starch Digestion

1. What was the result of adding the iodine solution to the starch solution?

2. What were the colors of the test tubes prior to heating?

3. How does the color change indicate that digestion has occurred?

4. What pancreatic enzyme is responsible for the breakdown of starch?

5. Where does digestion of carbohydrates take place?

6. What enzymes are responsible for digestion of starch?

7. Where does absorption of carbohydrates take place?

Part 3. Fat Digestion

1. What caused the color change in the test tube in test tube F?

2. Although there was no test for this product, what is the other product formed in the digestion of butter fat?

3. What pancreatic enzyme is responsible for the digestion of fat?

4. Where does the digestion of fat occur in the body?

5. In which type of vessels are fatty acids transported to the liver?

11
L A B O R A T O R Y

Anatomy of the Urinary System

A. Instructional Objectives: After completing this exercise the student should be able to:

1. Describe the function of the urinary system.
2. Identify, on a diagram or torso model, the urinary system organs and the function of each.
3. Compare the lengths of the male and female urethra.
4. Identify the following regions of the kidney model and/or sheep kidney:
   a. Hilus
   b. Cortex
   c. Medulla
   d. Medullary pyramids
   e. Major/minor calyces
   f. Pelvis
   g. Renal columns
   h. Renal capsule
5. Trace the blood supply of the kidney from the abdominal aorta back to the vena cava.
6. Identify a nephron, its function (s) and its parts on models: glomerulus, Bowman's capsule, proximal convoluted tubule, Loop of Henle, distal convoluted tubule and collecting duct.
7. Define glomerular filtration, tubular reabsorption, and tubular secretion and identify what parts of the nephron they happen in.
8. Define micturition.
9. Recognize microscopic views of the histologic structure of the kidney (i.e., glomerulus and Bowman's capsule, renal tubule, nephron).
10. List the physical characteristics of urine.
11. Indicate the normal pH and specific gravity ranges.
12. List substances that are normally found in the urine.
13. List substances that are abnormal urinary constituents.

14. Conduct a urinalysis test and use it to determine what substances are present in the urine.
15. Define the following terms and explain the implications and possible causes of the following conditions:
    a. Albuminuria
    b. Calculi
    c. Glycosuria
    d. Hematuria
    e. Hemoglobinuria
    f. Ketonuria
    g. Pyuria

B. Materials
    1. Human dissectible torso model
    2. Three-dimensional model of urinary system
    3. Anatomical chart of urinary system
    4. Sheep kidney
    5. Dissecting instruments and tray
    6. Three-dimensional models of cut kidney and of a nephron
    7. Compound microscope
    8. Prepared slides of a longitudinal and cross section of kidney
    9. Numbered artificial specimens provided by instructor
    10. Dipsticks provided by instructor
    11. Urine specimen cup
    12. Disposable gloves

C. Introduction

When the body metabolizes nutrients, it produces wastes such as carbon dioxide, nitrogenous wastes, and ammonia. The urinary system functions to eliminate these wastes from the body. Although some other organ systems also remove wastes from the body (e.g., lungs remove carbon dioxide), the urinary system is primarily concerned with removal of nitrogenous wastes. It also helps maintain homeostasis by maintaining the electrolyte, acid-base, and fluid balances of the blood.

The kidneys act to filter and process the blood. They allow the body to keep the necessary substances in the blood while allowing unnecessary substances such as toxins, metabolic wastes, and excess ions to leave the body through the urine. Without the urinary system, the body cannot maintain homeostasis. Without homeostasis, organ failure will occur. If homeostasis is not returned, it could result in death

Blood composition depends on three major factors: diet, cellular metabolism, and urinary output. In 24 hours, the kidneys' 2 million nephrons filter approximately 150 to 180 liters of blood plasma through their glomeruli into the tubules, where it is selectively processed by tubular reabsorption and secretion. In the same period, urinary output, which contains by-products of metabolism and excess ions, is 1.0 to 1.8 liters. In healthy individuals, the kidneys can maintain blood constancy despite wide variations in diet and metabolic activity. With certain pathological conditions, urine composition often changes dramatically.

D. Methods
    1. Gross anatomy of the human urinary system

       The urinary system (Figure 11.1a) includes a pair of **kidneys** and ureters and one urinary bladder and urethra. The kidneys perform the functions described previously and manufacture urine in the process. The ureters, bladder, and urethra either store the urine temporarily or help transport the urine out of the body.

    2. Identifying urinary system organs

       Examine the human torso model or a three-dimensional model of the urinary system to locate and study the anatomy and relationships of the urinary organs.

a. Locate the pair of kidneys on the dorsal body wall in the superior lumbar region. Notice that the right kidney is slightly lower than the left kidney. This is because the liver crowds the right kidney. See Figure 11.1b. In a living person, the kidneys are held in place by fat deposits and renal capsules.

b. If the amount of fat around the kidney is reduced, the kidneys will not be as securely anchored against the body wall. They may drop to a lower position in the abdominal cavity. This phenomenon is called **ptosis**.

c. Also, notice the relationship of the kidneys to the other organs in the abdominal cavity (Figure 11.1c)

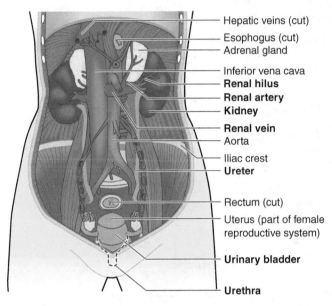

Hepatic veins (cut)
Esophogus (cut)
Adrenal gland
Inferior vena cava
Renal hilus
Renal artery
Kidney
Renal vein
Aorta
Iliac crest
Ureter
Rectum (cut)
Uterus (part of female reproductive system)
Urinary bladder
Urethra

FIGURE 11.1A   Anterior view of urinary system organs

Figure 40.1a from *Marieb's Human Anatomy and Physiology Laboratory Manual*, 8th edition

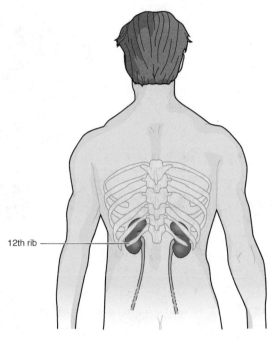

12th rib

FIGURE 11.1B   Posterior view showing position of kidneys.

Figure 40.1b from *Marieb's Human Anatomy and Physiology Laboratory Manual*, 8th edition

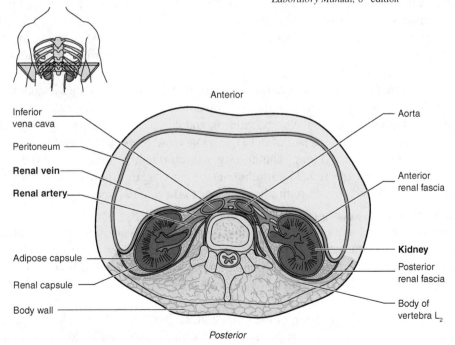

Anterior

Inferior vena cava
Peritoneum
Renal vein
Renal artery
Adipose capsule
Renal capsule
Body wall

Aorta
Anterior renal fascia
Kidney
Posterior renal fascia
Body of vertebra L$_2$

Posterior

FIGURE 11.1C   Transverse section of abdomen

Figure 40.1c from *Marieb's Human Anatomy and Physiology Laboratory Manual*, 8th edition

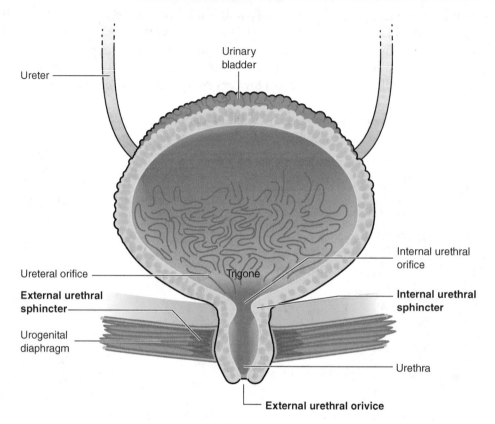

FIGURE 11.2   Detailed structure of urinary bladder, urethra sphincters, and urethra of female.
Figure 40.2 from *Marieb's Human Anatomy and Physiology Laboratory Manual,* 8th edition

    d. The **renal arteries** branch from the descending aorta and enter the hilus of the kidney. The **renal veins** drain the kidneys of blood and send the blood back to the vena cava. The **ureters** drain urine from the kidneys and send it by peristalsis to the bladder for temporary storage. Figure 11.3 illustrates the internal anatomy of the kidney.

    e. Locate the **urinary bladder** and notice where the ureters enter it. Also locate the urethra, which drains the bladder. The two ureteral and one urethral openings form a triangular region on the bladder. This region is called the **trigone**.

    f. Follow the urethra from the bladder to the body exterior. The male's urethra is approximately 20 cm (8 inches) long. It travels the length of the **penis** and opens at the tip. The male urethra has two functions: it allows urine to pass to the body exterior, and it gives semen a passageway out of the body. Thus, in the male, the urethra is part of both the urinary and reproductive systems. The female's urethra is only 4 cm (1.5 inches) long. The urinary and reproductive structures of the female are completely separate. Thus, the female's urethra serves only to transport urine out of the body. The external opening of the female urethra lies anterior to the vaginal opening.

3. Functional microscopic anatomy of the kidney and bladder

The functional anatomical unit of the kidney is the **nephron**. There are more than 1 million nephrons in each kidney. The structure of the nephron and its location in the kidney are illustrated in Figure 11.4.

The nephron is made of two major structures: a **glomerulus** and a renal tubule. The glomerulus is essentially a ball of capillaries. It is surrounded by a **glomerular (Bowman's) capsule**. The glomerulus and Bowman's capsule together are called the **renal corpuscle**.

The Bowman's capsule is the first part of the renal tubule. The rest of the tubule is approximately 3 cm (1.25 inches) long. As it emerges from the glomerular capsule, it becomes highly coiled and convoluted, drops down into a long hairpin loop, and then again coils and twists before entering a collecting duct. In

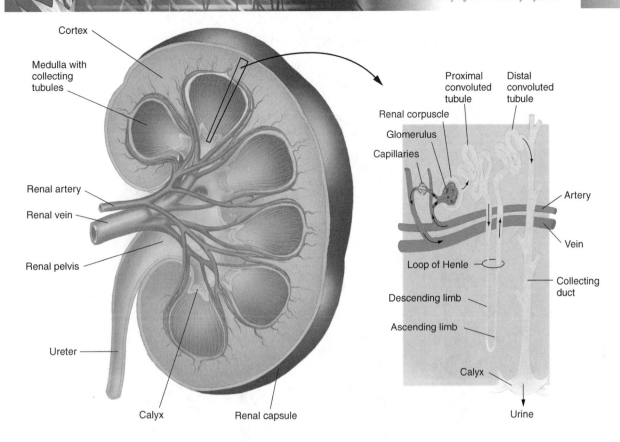

FIGURE 11.3   Frontal section of a kidney

Figure 3.1 from *Marieb's Human Anatomy and Physiology Laboratory Manual,* 8th edition

order from the glomerular capsule, the anatomical areas of the renal tubule are as follows: the proximal convoluted tubule, loops of Henle (descending and ascending limbs), and the distal convoluted tubule. The wall of the renal tubule is composed almost entirely of cuboidal epithelial cells, with the exception of part of the descending limb of the loop of Henle, which is simple squamous epithelium.

Most nephrons, called cortical nephrons, are located entirely within the cortex. However, parts of the loops of Henle of the juxtamedullary nephrons (close to the cortex-medulla junction) penetrate well into the medulla. The collecting ducts, each of which receives urine from many nephrons, run downward through the medullary pyramids, giving them their striped appearance. As the collecting ducts approach the renal pelvis, the urine drains into the pelvis of the kidney.

The function of the nephron depends on several unique features of the renal circulation. The capillary vascular supply consists of two distinct capillary beds: the glomerulus and the peritubular capillary bed. Vessels leading to and from the glomerulus, the first capillary bed, are both arterioles: the afferent arteriole feeds the bed while the efferent arteriole drains it. It is a high-pressure bed along its entire length. The high hydrostatic pressure created by these two anatomical features forces out fluid and small blood components from the glomerulus into the glomerular capsule. That is, it forms the filtrate, which is processed by the nephron tubule.

The peritubular capillary bed arises from the efferent arteriole draining the glomerulus. This set of capillaries clings intimately to the renal tubule and empties into the veins of the kidney. The peritubular capillaries are low-pressure porous capillaries adapted for absorption rather than filtration and readily take up the solutes and water reabsorbed from the filtrate by the tubule cells. The glomerulus produces the filtrate and the peritubular reclaim most of that filtrate.

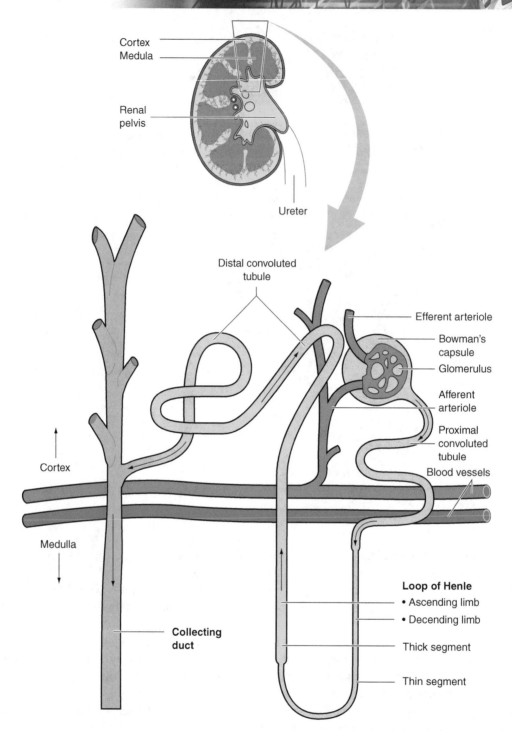

FIGURE 11.4   Structure of a nephron.

From *Marieb's Human Anatomy and Physiology Laboratory Manual,* 8th edition figure 40.4

Urine formation is a result of three processes: filtration, reabsorption, and secretion. **Filtration**, the role of the glomerulus, is largely a passive process in which a portion of the blood passes from the glomerular bed into the glomerular capsule. This filtrate then enters the proximal convoluted tubule, where tubular **reabsorption** and secretion begin. During tubular reabsorption, many of the filtrate components move through the tubule cells and return to the blood in the peritubular capillaries. Some of this reabsorption is passive, such as that of water, which passes by osmosis, but the reabsorption of most substances depends on active transport processes and is highly selective. Which substances are reabsorbed at a particular time depends on the composition of the blood and needs of the body at that time. Substances that are almost entirely reabsorbed from the filtrate include water, glucose, and amino acids. Various ions are selectively reabsorbed or allowed to go out in the urine according to what is required to maintain appropriate blood pH and electrolyte composition. Waste products (urea, creatinine, uric acid, and drug metabolites) are reabsorbed to a much lesser degree or not at all. Most (75% to 80%) of tubular reabsorption occurs in the proximal convoluted tubule. The balance occurs in other areas, especially the distal convoluted tubules and collecting ducts.

Tubular secretion is essentially the reverse process of tubular reabsorption. Substances such as hydrogen ions, potassium ions, and creatinine move either from the blood of the peritubular capillaries through the tubular cells or from the tubular cells into the filtrate to be disposed of in the urine. This process is particularly important for the disposal of substances not already in the filtrate (such as drug metabolites) and as a device for controlling blood pH.

4. Studying nephron structure

Begin your study of nephron structure by identifying the glomerular capsule, proximal and distal convoluted tubule regions, and the loop of Henle on a model of the nephron. Then, obtain a compound microscope and a prepared slide of kidney tissue to continue with the microscope study of the kidney.

Secure the slide on the microscope stage, and scan the slide under low power.

Identify a glomerulus, which appears as a ball of tightly packed material containing many small nuclei.

5. Bladder

Although urine production by the kidney is a continuous process, urine is usually removed from the body when voiding is convenient. In the meantime the urinary bladder, which receives urine via the ureters and discharges it via the urethra, stores it temporarily.

Voiding, or micturition, is the process in which urine empties from the bladder. Two sphincter muscles or valves, the internal urethral sphincter and external urethral sphincter, control the outflow of urine from the bladder. Ordinarily, the bladder continues to collect urine until about 200 mL has accumulated, at which time the stretching of the bladder wall activates stretch receptors. As contractions increase in force and frequency, stored urine is forced past the internal sphincter, which is a smooth muscle involuntary sphincter, into the superior part of the urethra. It is then that a person feels the urge to void. The inferior external sphincter consists of skeletal muscle and is voluntarily controlled. If it is not convenient to void, the opening of this sphincter can be inhibited. Conversely, if the time is convenient, the sphincter may be

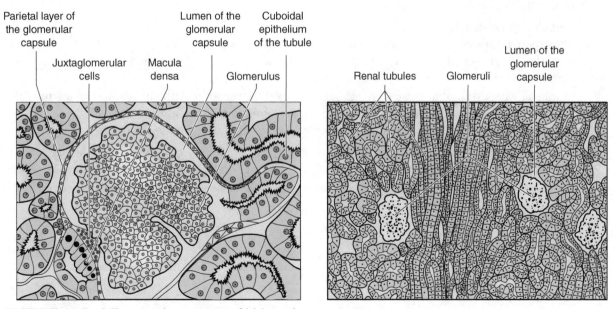

FIGURE 11.5   Microscopic structure of kidney tissue

From Figuer 40.6 in *Marieb's Human Anatomy and Physiology Laboratory Manual,* 8th edition.

relaxed and the stored urine flushed from the body. If voiding is inhibited, the reflex contractions of the bladder cease temporarily and urine continues to accumulate in the bladder. After another 200 to 300 mL of urine has been collected, the micturition reflex will again be initiated.

Lack of voluntary control over the external sphincter is referred to as incontinence. Incontinence is normal in children age two years and younger, as they have not yet gained control over the voluntary sphincter. In adults and older children, incontinence is generally a result of spinal cord injury, emotional problems, bladder irritability, or some other pathology of the urinary tract.

6. Characteristics of urine

To be valuable as a diagnostic tool, a urinalysis must be done within 30 minutes after the urine is voided or the urine must be refrigerated. Freshly voided urine is generally clear and pale yellow to amber in color. As a rule, color variations from pale yellow to deeper amber indicate the relative concentration of solutes to water in the urine. The greater the solute concentration, the deeper the color. Abnormal urine color may be due to certain foods, such as beets, various drugs, bile, or blood.

The odor of freshly voided urine is slightly aromatic, but bacterial action gives it an ammonia-like odor when left standing. Some drugs, vegetables (e.g., asparagus), and various disease processes (e.g., diabetes mellitus) alter the characteristic odor of urine. For example, the urine of a person with uncontrolled diabetes mellitus (and elevated levels of ketones) smells fruity or acetone-like.

The pH of urine ranges from 4.5 to 8.0, but its average value, 6.0, is slightly acidic. Diet may markedly influence the pH of the urine. For example, a diet high in protein (meat, eggs, cheese) and whole wheat products increases the acidity of urine. Such foods are called acid ash foods. On the other hand, a vegetarian diet (alkaline ash diet) increases the alkalinity of the urine. A bacterial infection of the urinary tract may also result in urine with a high pH.

Specific gravity is the relative weight of a specific volume of liquid compared with an equal volume of distilled water. The specific gravity of distilled water is 1.000. Since urine contains dissolved solutes, it weighs more than water, and its customary specific gravity of 1.001 to 1.030 contains few solutes and is considered very dilute. Dilute urine commonly results when a person drinks excessive amounts of water, uses diuretics, or suffers from diabetes insipidus or chronic renal failure. If urine becomes excessively concentrated, some of the substances normally held in solution begin to precipitate or crystallize, forming kidney stones, or renal calculi.

Normal constituents of urine include water, urea, sodium, potassium, phosphate ions, sulfate ions, creatinine, and uric acid. Much smaller but highly variable amounts of calcium, magnesium, and bicarbonate ions are also found in the urine. Abnormally high concentrations of any of these urinary constituents may indicate a pathological condition.

7. Analyzing urine samples

Use prepared dipsticks and perform chemical tests to determine the characteristics of normal urine as well as to identify abnormal urinary components. Investigate both normal urine and unknown urine specimens provided by the instructor. Record results on Chart 11.1.

Obtain and wear disposable gloves throughout this laboratory session. Although the instructor-provided urine samples are actually artificial urine (concocted in the laboratory to resemble real urine), the techniques of safe handling of body fluids should still be observed as part of your learning process.

8. Determination of the physical characteristics of urine

Determine the color, transparency, and odor of your "normal" sample and unknown samples.

Record your results in the chart.

Table 11.1 Abnormal Urinary Constituents

Constituent	Condition	Indications
Glucose	**Glycosuria**	Abnormally high blood sugar levels **Nonpathological:** High carbohydrate intake Pathological: Uncontrolled diabetes mellitus
Albumin	**Albuminuria**	Abnormally increased permeability of glomerulus **Nonpathological:** Excessive exertion, pregnancy, overabundant protein intake **Pathological:** Damage to glomerular membrane; for example, kidney trauma, ingestion of poison or metal, bacterial toxins, hypertension, glomerulonephritis
Ketone bodies	**Ketonuria**	**Pathological:** Starvation, diets low in carbohydrates, inadequate food intake, all of which force the body to use fat stores
Red blood cells	**Hematuria**	**Nonpathological:** Menstruating female **Pathological:** Irritation of urinary tract organs (by kidney stones), infection or tumors of urinary tract, physical trauma to urinary organs
Hemoglobin	**Hemoglobinuria**	Fragmentation of red blood cells **Pathological:** Hemolytic anemia, transfusion reaction, burn, poisonous snake bites, renal disease
Nitrites	**Nitrites**	**Pathological:** Bacterial infection
Bile pigments	**Bilirubinuria**	**Pathological:** Liver; for example, hepatitis, cirrhosis, bile duct blockage
White blood cells	**Pyuria**	**Pathological:** Inflammation of urinary tract

CHART 11.1   Urinalysis Results

Observation or test	Normal values	Standard urine specimen	A	B	C
Physical characteristics					
Color	Pale yellow	yellow: pale, medium, dark			
Transparency	Transparent	clear, slightly cloudy, cloudy			
Odor	Characteristic	___	___	___	___
pH	4.5–8.0	___	___	___	___
Specific gravity	1.010–1.030	___	___	___	___
Inorganic components					
Nitrites	Negative	___	___	___	___
Organic components					
Urea	Present	___	___	___	___
Glucose	Negative	___	___	___	___
Protein	Negative	___	___	___	___
Ketone bodies	Negative	___	___	___	___
Blood/hemoglobin	Negative	___	___	___	___
Bilirubin	Negative	___	___	___	___
Urobilinogen	Present	___	___	___	___
Leukocytes	Negative	___	___	___	___

CHART 11.2　Factors Affecting Urinalysis Results

Observation or test	Normal values	Factors
Physical characteristics		
Color	pale yellow	
Transparency	transparent	
Odor	characteristic	
pH	4.5–8.0	
Specific gravity	1.010–1.030	
Inorganic components		
Nitrites	Negative	
Organic components		
Urea	Present	
Glucose	Negative	
Protein	Negative	
Ketone bodies	Negative	
Blood/hemoglobin	Negative	
Bilirubin	Negative	
Leukocytes	Negative	

5. Questions

Explain the significance of the following terms:

Glycosuria

Albuminuria

Ketonuria

Hematuria

Hemoglobinuria

Pyuria

12 LABORATORY

The Skeletal System

A. Instructional Objectives: After completing this exercise, the student should be able to:
1. List five functions of the skeletal system.
2. Distinguish between the axial and appendicular skeleton.
3. Recognize and give the name and location of the bones of the axial skeletal system.
4. Recognize and give the name and location of the bones of the appendicular skeletal system.
5. Recognize and give the name and location of any bone listed in this exercise.
6. Classify any bone as being long, flat, short or irregular.
7. Identify the following parts of a bone: diaphysis, marrow or medullary cavity, yellow marrow, red marrow, and epiphysis.
8. List the location of cartilage in the body.
9. Locate and identify examples of the following types of joints
   a. Immovable
   b. Partially movable
   c. Freely movable
      (1) Ball and socket
      (2) Hinge
      (3) Pivot
10. Define all terms listed in this exercise, as well as tendon, ligament, compact bone, spongy bone, joint, articulation, suture, calcium, endoskeleton, and exoskeleton.
11. Complete the questions at the end of this exercise.

B. Introduction

The skeletal system performs five important functions: (1) support, (2) protection, (3) attachment for muscles for movement, (4) production of blood cells, and (5) calcium storage and release. Without support, your body would have no form or shape to it. The skeletal system gives your body structure. Certain animals, such as fish, dogs, cats, and other mammals have an internal skeleton or endoskeleton. Lower animals such as arthropods and echinoderms have an external skeleton or exoskeleton. Regardless of the type of skeleton, however, the internal organs are protected. For example, our rib cage protects our heart and lungs. Muscles are usually attached to two bones by a tendon to facilitate motion. Bones are attached to one another by ligaments. Red blood cells, white blood cells, and platelets are produced in the red marrow portion of the bones. These cells circulate in your bloodstream to offer many benefits. Bones are composed of calcium compounds and thus serve as a reservoir for calcium as it is needed to perform its many physiological functions in the body.

There are approximately 206 bones in the human body. Recognizing and memorizing them is a fun yet challenging task. Try to locate each bone and joint on the skeletons in the lab and on your own body.

C. Materials

Complete human skeletons

Human skulls

Human leg bones

Human arm bones

Vertebral column, assembled

Loose vertebra

D. Methods

The instructor will pronounce the names of the bones with the class. You are responsible for locating each bone on the skeletons and labeling these structures on the drawings that follow. You should be able to identify all the bones before you leave the lab.

The skeleton is divided into the axial skeleton and appendicular skeleton.

Axial Skeleton

1. Skull (cranium)

   The skull is composed of 28 separate bones that make up the brain case, the face, and the middle ear. Most of the bones articulate with one another through immovable joints called sutures. Only the more prominent bones will be covered in this exercise.

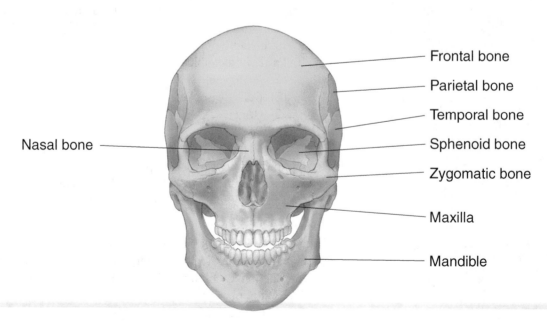

FIGURE 12.1   Anterior view of skull

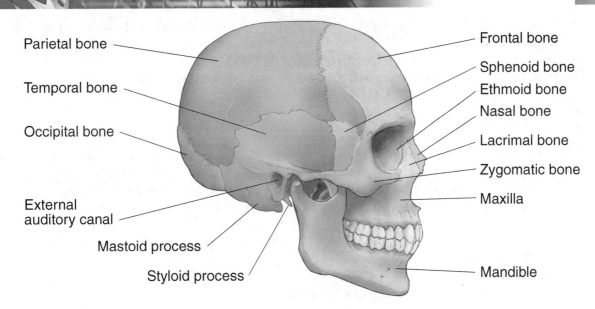

Parietal bone

Temporal bone

Occipital bone

External
auditory canal

Mastoid process

Styloid process

Frontal bone

Sphenoid bone

Ethmoid bone

Nasal bone

Lacrimal bone

Zygomatic bone

Maxilla

Mandible

FIGURE 12.2   Lateral view of skull

Copyright © Kendall/Hunt Publishing Company.

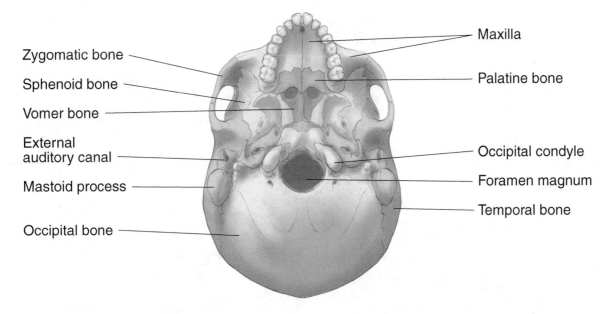

Zygomatic bone

Sphenoid bone

Vomer bone

External
auditory canal

Mastoid process

Occipital bone

Maxilla

Palatine bone

Occipital condyle

Foramen magnum

Temporal bone

FIGURE 12.3   Inferior view of skull

Copyright © Kendall/Hunt Publishing Company.

a. **Occipital**—single bone making up the lower back part of the head
b. **Frontal**—single bone forming the forehead
c. **Parietal**—paired bones just anterior to the occipital
d. **Temporal**—paired bones just inferior to the parietals; make up posterior portion of the zygomatic arch
e. **Sphenoid**—single bone forming much of the floor of the brain case
f. **Ethmoid**—single bone forming much of the internal structure of the nose; locate parts of it as projections in the upper areas inside the nose; this bone, along with the vomer, also partitions the nasal passage
g. **Maxillae**—paired bones of the upper jaw
h. **Nasal (2)**—paired bones that constitute the bony portion of the external nose

i. **Lacrimal**—small paired bones in the medial wall of the eye socket behind the nasal and upper part of the maxillae

j. **Zygomatic**—paired bones that form front part of the cheek

k. **Palatine**—paired bones forming posterior portion of the roof of the mouth

l. **Mandible**—lower jaw

m. **Auditory ossicles**

2. **Vertebral column**

The vertebral column consists of 26 vertebrae and the joints between them, most of which are partially movable. The first two bones are given individual names. The atlas is the first vertebra and holds up the head. The axis is the second vertebra and is located just under the atlas. The vertebrae form a continuous housing for the spinal cord. There are five regions of vertebrae.

a. **Cervical vertebrae (7)**—neck

C1-C7

    (1)   Atlas (C1)

    (2)   Axis (C2)

b. **Thoracic vertebrae (12)**—chest; attach to ribs

T1-T12

c. **Lumbar vertebrae (5)**—lower back

L1-L5

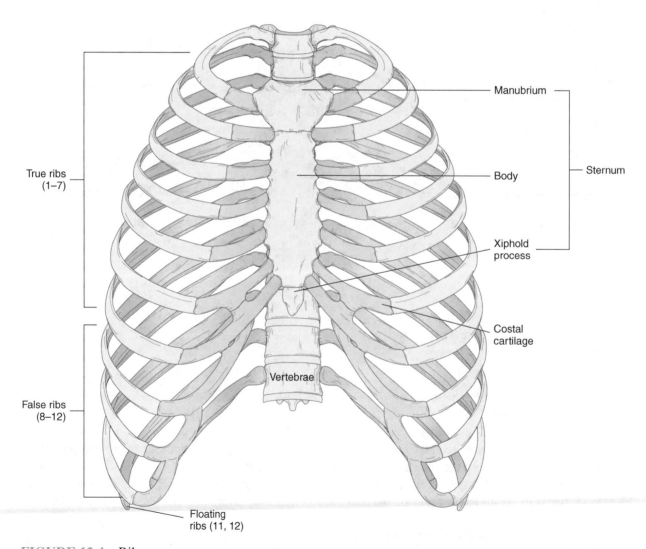

FIGURE 12.4   Rib cage. © *Kendall/Hunt Publishing Company.*

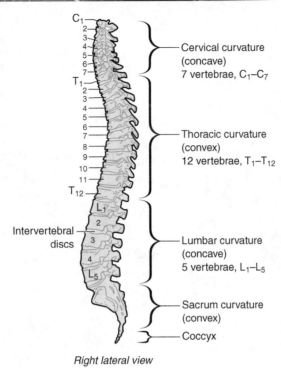

C₁
2
3
4 — Cervical curvature
5   (concave)
6   7 vertebrae, C₁–C₇
7
T₁
2
3
4
5
6
7 — Thoracic curvature
8   (convex)
9   12 vertebrae, T₁–T₁₂
10
11
T₁₂
L₁
2 — Lumbar curvature
Intervertebral 3   (concave)
discs 4   5 vertebrae, L₁–L₅
L₅
— Sacrum curvature
   (convex)
— Coccyx

Right lateral view

FIGURE 12.5 Divisions of vertebral column

     d. **Sacrum**—several fused vertebrae below the lumbar and between the hips
     e. **Coccyx**—several fused vertebrae below the sacrum; tailbone

3. **Rib cage**

The rib cage consists of 12 pairs of **ribs** and the **sternum** or breastbone. The first seven pairs are **true ribs** because they are connected directly to the sternum. The other five pairs are called **false ribs** because they are indirectly connected to the sternum. The last two pairs of false ribs are also called **floating ribs** because they are not attached either directly or indirectly to the sternum.

     a. **Ribs**
        (1) **True**
        (2) **False**
           i. **Floating**
     b. **Sternum**

Appendicular Skeleton

4. **Upper extremity (shoulders and arms)**—consists of 32 pairs of bones, each of which will be considered individually
     a. Shoulder
        (1) **Clavicle**—collarbone
        (2) **Scapula**—shoulder blade
     b. Upper appendage
        (1) **Humerus**—long bone of the upper arm; joint between the scapula and the humerus is of the freely movable ball-and-socket type
        (2) **Radius**—long bone of the lower arm that extends to the thumb side of the hand
        (3) **Ulna**—long bone of the lower arm that extends to the little-finger side of the hand; articulates with the humerus to form the elbow, a freely movable hinge type of joint
        (4) **Carpals**—the eight short bones in the wrist
        (5) **Metacarpals**—the five bones of each hand
        (6) **Phalanges**—the 14 bones of the fingers

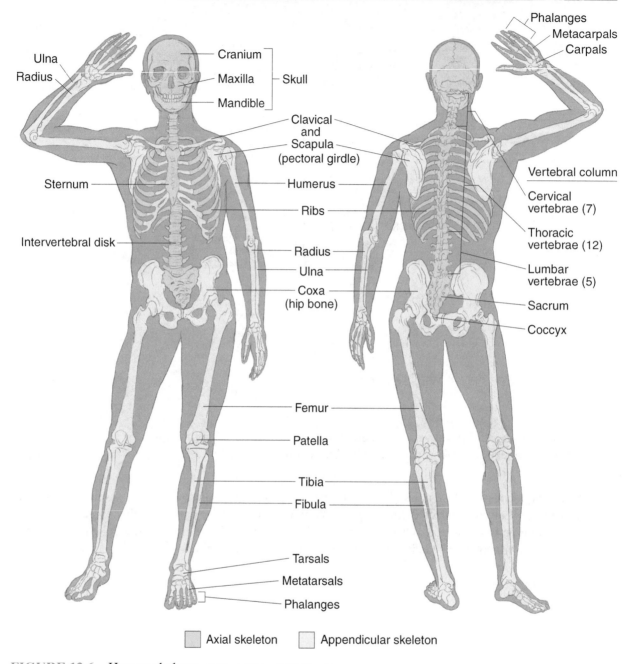

FIGURE 12.6   Human skeleton. © *Kendall/Hunt Publishing Company.*

5. Lower extremity (hips and legs)—consists of 31 pairs of bones, each of which will be considered individually
   a. Pelvis
      (1) Coxal (2)—Large bone on each side of the body. The bony structure of the pelvis is quite different in males and females. In females the opening between the two coxal bones and in front of the sacrum and coccyx is considerably larger to facilitate childbirth. Each coxal bone contains a hip socket.
   b. Lower appendage
      (1) Femur—thigh bone; longest bone of the body
      (2) Patella—kneecap
      (3) Tibia—larger of the two leg bones

(4) **Fibula**—smaller of the two leg bones

(5) **Tarsals**—the seven short bones of each ankle; includes the heel bone

(6) **Metatarsals**—the five bones in each foot

(7) **Phalanges**—the 14 bones in the toes of each foot

6. Hyoid bone

7. Classification of bones

Bones may be classified by their shape. The four shapes are long, flat, short, and irregular. The major bones of the arms and legs are called **long bones**. The bones of the rib cage and the skull case are classified as flat bones. Those of the wrist and ankle are the **short bones**. The vertebrae and a few bones of the cranium are called **irregular bones** because they do not fit one of the other categories.

8. Structure of bones

With few exceptions, bones have a hardened layer of **compact bone** on the periphery with a center of porous or **spongy bone**. The compact bone gives strength of rigidity to the skeleton, while the spongy bone permits lightness. The skeleton would weigh twice as much if it were made up totally of compact bone. The articular cartilage is made of hyaline cartilage and covers surfaces of bones at points of articulation with other bones.

Long bones are constructed differently from the others. The shaft of the bone is the **diaphysis**; it is cylindrical and made of compact bone. The hollow center is the marrow cavity or **medullary canal** and is filled with stored fat called **yellow marrow**. Each long bone is flared at the ends. This terminal enlargement is called the **epiphysis** and is made principally of spongy bone with a thin shell of compact bone. The **epiphyseal disc** is a cartilaginous part of the between located between the diaphysis

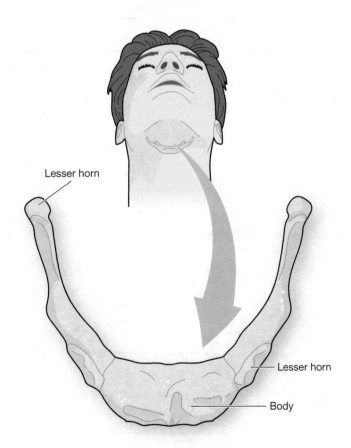

Lesser horn

Lesser horn

Body

FIGURE 12.7   Hyoid bone. From *Human Anatomy and Physiology Laboratory Manual*. Cat Version. 8th ed. by Elaine N. Marieb. Copyright © 2005 by Pearson Education, Inc. Reprinted by permission.

and the epiphysis. This is the "growth plate" where the long bone grows in length. The porous spaces within the spongy bone is filled with red marrow, which is where blood cells are manufactured.

9. Joints of the body

There are three types of joints in the body: immovable, partially movable, and freely movable. The best known immovable joints are the sutures of the skull. Partially movable joints may be found in the intervertebral discs and the pubic symphysis.

The three major types of freely movable joints are the ball-and-socket, hinge, and pivot joint. The ball-and-socket joints allow movement in all directions. They may be found in the hip and shoulder. Hinge joints allow movement in only one direction. These joints are found in the knee, elbow, phalanges, and jaw. The pivot joint is found between the atlas and the axis and between the radius and the ulna. The atlas pivots around the odontoid process of the axis to allow the skull to turn to the right and left.

10. Cartilage

Cartilage is found in several places in the adult human skeleton and makes up the entire skeleton of the fetus while in the mother's womb. Look for cartilage on the skeleton in the following locations: (1) external ear, (2) nose, (3) intervertebral discs, (4) attaching the ribs to the sternum, and (5) joints.

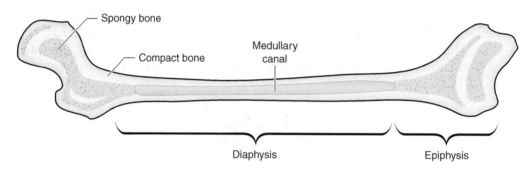

FIGURE 12.8   Structure of long bone

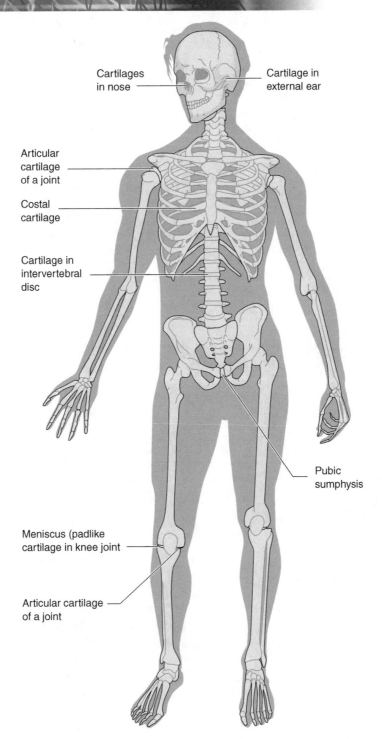

FIGURE 12.9 **Cartilages in the adult skeleton.** From *Human Anatomy and Physiology*, 8th by Elaine N. Marieb and Katja Hoehn. Copyright © 2010 by Pearson Education, Inc. Reprinted by permission.

Name: _____

QUESTIONS

1. List five functions of the skeletal system.

2. What type of joint is found in each of the places below?
   a. Between parietal and temporal bones
   b. Between atlas and axis
   c. Between the humerus and scapula
   d. Between the humerus and ulna
   e. Between two of the lumbar vertebrae
   f. Between the tibia and the femur

3. List four places where cartilage is found.

4. Classify each of the following bones as being long, short, flat, or irregular.
   a. Rib   _____
   b. Vertebra   _____
   c. Zygomatic   _____
   d. Occipital   _____
   e. Humerus   _____
   f. Tarsals   _____
   g. Fibula   _____

5. Classify each of the following bones as being part of the axial or appendicular skeleton.
   a. Ethmoid   _____
   b. Tarsals   _____
   c. Sacrum   _____
   d. Occipital   _____
   e. Phalanges   _____
   f. Ribs   _____
   g. Atlas   _____
   h. Radius   _____

13 LABORATORY

The Nervous System

A. Instructional Objectives

After completing this exercise, the student should be able to

1. Identify from a model or illustration the following parts of a motor neuron, and know their function:

Cell body	Nissl bodies
Nucleus	Axon
Dendrites	Myelin sheath
Nodes of Ranvier	Neurilemma

2. Identify from a model or illustration the following parts of the spinal cord in cross-section, and know their function: motor neuron, association neuron, dorsal root, dorsal root ganglion, ventral root, gray and white matter, receptor, effector, spinal nerve.

3. Identify the following structures from the model of the human brain, and know their functions:

   a. From the sagittal view—corpus callosum, thalamus, midbrain, occipital lobe, fourth ventricle, cerebellum, spinal cord, medulla, pons, temporal lobe, pituitary gland, hypothalamus, third ventricle, frontal lobe, cerebrum, parietal lobe.

   b. From the ventral view—olfactory bulb, optic chiasm, pituitary gland, temporal lobe, pons, medulla, cerebellum, spinal cord, occipital lobe, parietal lobe, midbrain, frontal lobe

   c. From the lateral view—central fissure, sensory area, lateral fissure, visual area, olfactory area, auditory area, speech area, motor area

4. Vocabulary—All terms used in this exercise, with the following additional terms: regeneration, central and peripheral nervous systems, cerebrospinal fluid, endocrine glands, motor activity.

From *Laboratory Exercises for an Introduction to Biological Principles* by Gil Desha. Reprinted by permission of the author.

B. Introduction

The central nervous system (CNS) in humans is composed of millions of cells, with millions of interconnections between the nerve cells. The interactions of the nerve cells with each other, with muscle cells, and with sensory receptors allow the human to learn, remember, feel, and react to various stimuli. The task of studying the nervous system can be staggering at first, but if the student takes one concept at a time, the knowledge is easier to understand and remember. Specific information for each nervous system area will be found in the Methods section of the exercise.

C. Materials

The laboratory should have plastic human brains and charts, spinal cord models and charts, plastic neuron models, and microscopes with the giant multipolar neuron smear slides.

D. Methods

Each student should work independently on the first two sections. Use the textbook, plastic neuron model, microscope with slide, and plastic spinal cord model.

1. The Neuron—The individual nerve cells are called neurons, and they serve as the basic functional unit of the nervous system. Their main function is to transmit electrical signals from one nerve cell to another, and from nerve cells to muscle cells. All neurons have an irregular shape, and consist of a cell body with numerous extensions of the plasma membrane called dendrites and one axon. Neurons are arranged so that the terminal axon branches lie very close to the dendrites or cell bodies of other neurons. The space between the two neurons is called a synaptic cleft, with the first neuron called the presynaptic neuron, and the second one called the postsynaptic neuron. The electrical impulse will travel along the dendrites and the axon, but will not cross the synaptic cleft. Instead, a chemical called a neurotransmitter will cross the space from the presynaptic neuron to the postsynaptic neuron. The numerous dendrites are processes that carry electrical signals towards the cell body of a neuron, and are sometimes called the input zone. The single, long axon is a process that carries electrical signals away from the cell body, and is sometimes called the output zone. The cell body contains a prominent nucleus in its center, surrounded by cytoplasm in which the typical cellular organelles are suspended. There are two critical differences, however, between neurons and cell bodies of other body cells. One is that the neuron cannot undergo mitosis, thus does not continually divide to replace or repair itself. It does not contain a centriole, or a centrosome—these organelles are needed for the formation of the spindle apparatus, but not in a cell with no mitotic potential. The second critical difference is that the neuron must retain the ability to manufacture proteins, but since it cannot divide, the rough endoplasmic reticulum has evolved into specialized structures called Nissl Bodies. The axon has a central core, or axis cylinder, surrounded by multiple layers of plasma membrane wrapped around it, called the myelin sheath. Some, not all, axons have a myelin sheath, but it is punctuated by gaps in the tissue, along the length of the axon. These gaps, called the nodes of Ranvier, allow the passage of ions to and from the axon and the surrounding aqueous material. The function of the myelin sheath is protection and insulation of the axon. Electrical signals do not pass through the myelin sheath, because the fatty material does not allow the passage of ions. The function of the nodes of Ranvier is to allow for the passage of ions, and therefore, the transmission of electrical signals, along the length of the axon. The electrical signal can travel in an extremely rapid manner by literally "jumping" over the myelin sheath, as it jumps from one node to the next. This method of transmission is called saltatory conduction. A thin membrane covering the myelin sheath of axons in the peripheral nervous system is called the neurilemma. It is needed if regeneration of injured nerve tissue is to occur. Study Figure 13.1, then label all the indicated parts in Figure 13.2.

2. The Spinal Cord and Nerve Pathways—There are three types of neurons associated with the formation of nerve pathways. Sensory (afferent) neurons transmit impulses from the skin, eyes, ears, mouth, nose, etc, to the spinal cord and brain. Receptors are proteins, functioning as sense organs, and are found on the surface of cells. Motor (efferent) neurons transmit impulses away from the brain and spinal cord to muscle and glands, known as effectors. Association neurons are found in the central nervous system, and form links between afferent and efferent neurons. Look at Figure 13.3, and study the way in which a nerve impulse passes through the spinal cord. Label the indicated

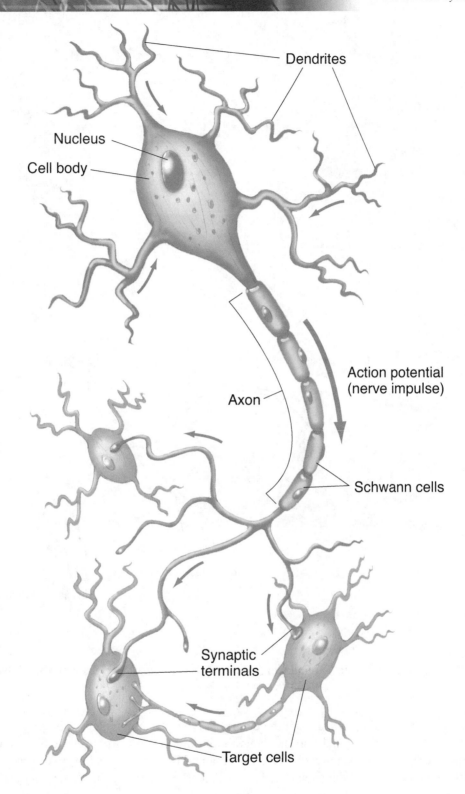

Dendrites

Nucleus

Cell body

Axon

Action potential
(nerve impulse)

Schwann cells

Synaptic
terminals

Target cells

FIGURE 13.1   Nerve Cell
Copyright © Kendall/Hunt Publishing Company.

structures: Notice that the cell bodies of sensory neurons lie outside of the spinal cord in the **dorsal root ganglia** (a ganglion is a collection of nerve cell bodies outside the CNS). The cell bodies of the association and motor neurons lie within the CNS, and make up the **gray matter** of the spinal cord. The **white matter** of the spinal cord is composed of ascending and descending nerve pathways (axonal tracts) to and from the brain. Nerves are composed of bundles of neuronal processes (axons

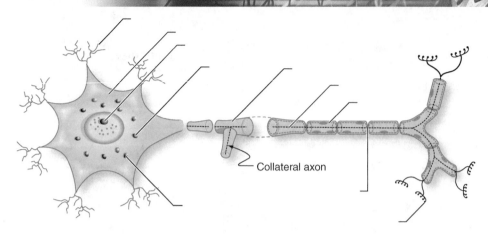

FIGURE 13.2   A Motor Neutron

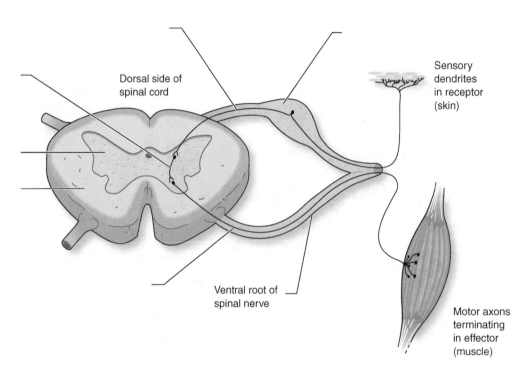

FIGURE 13.3   Spinal Nerve and Spinal Cord Showing Locations of Various Neuron Types

and dendrites) leading to and from the central nervous system, which is the brain and spinal cord. Nerves that enter the spinal cord at intervals through openings between the vertebrae are called spinal nerves. The human nervous system contains 31 pairs of spinal nerves. Nerves that enter the CNS above the level of the spinal cord are called cranial nerves, and the cranial and spinal nerves together constitute the Peripheral Nervous System.

3. The Mammalian Brain—The most complex organ of the body is composed of an extremely large number of association neurons and nerve fiber tracts. The structures and functions of the human brain are studied by using a laboratory model. Each pair of students should have a plastic laboratory model of the human brain. To start, refer to Figures 13.4, 13.5 and 13.6 and locate the following structures.

a. From the dorsal (Top) view—Find the cerebrum, the largest portion of the brain. It consists of 2 cerebral hemispheres, right and left, divided by a longitudinal fissure. The surface is covered

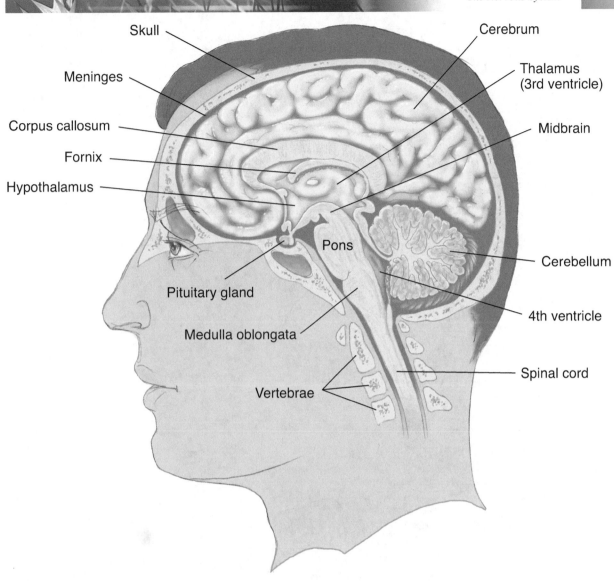

Skull

Cerebrum

Meninges

Thalamus
(3rd ventricle)

Corpus callosum

Midbrain

Fornix

Hypothalamus

Pons

Cerebellum

Pituitary gland

4th ventricle

Medulla oblongata

Spinal cord

Vertebrae

FIGURE 13.4 The Human Brain
Copyright © Kendall/Hunt Publishing Company.

with inward folds, called sulci, and outward bulges, called gyri. A very deep sulcus is a fissure. The transverse fissure is running in a horizontal direction, and separates the 2 cerebral hemispheres from the cerebellum. The cerebral surface (cortex) is covered by thin membranes called meninges. Each hemisphere is divided into lobes by prominent sulci. The names of the lobes correspond to bones of the skull. For example, the frontal lobe is the anterior portion of the brain, and corresponds to the frontal bone, which we typically think of as forming the forehead. The frontal lobe constitutes approximately 1/3 of the dorsal surface of each hemisphere. A central fissure separates the frontal lobe from the parietal lobe. This fissure is also known as the fissure of Rolando. The parietal lobe extends posteriorly from the central fissure to a small sulcus near the posterior end of the cerebrum. The occipital lobe is the smallest lobe, located posterior to the parietal lobe at the back of the cerebrum. The cerebellum is the part of the brain situated posterior, and inferior to the cerebrum. It consists of a pair of lateral lobes and a median lobe, called the vermis. The medulla oblongata is the most posterior portion of the brain stem. It resembles the spinal cord, but the diameter is thicker. Many nerve centers which give rise to the cranial

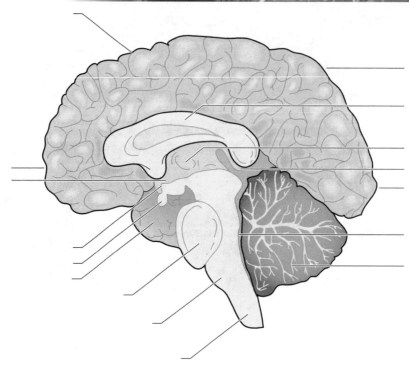

FIGURE 13.5   Midsagittal View of Human Brain

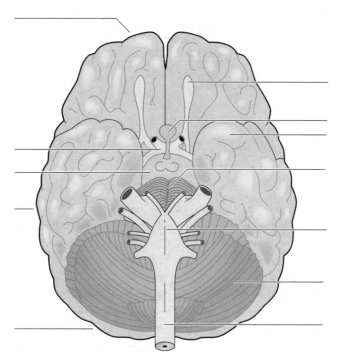

FIGURE 13.6   Ventral View of Human Brain
From *Laboratory Exercises for an Introduction to Biological Principles* by Gil Desha. Reprinted by permission of the author.

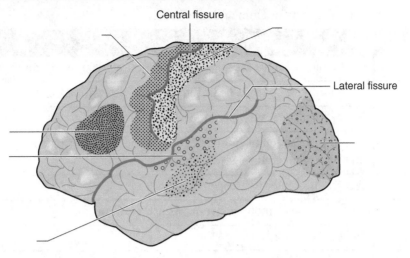

Central fissure

Lateral fissure

FIGURE 13.7   Lateral View of Human Brain Showing Areas of Cerebrum

nerves are found here. The midbrain is difficult to see—spread the cerebellum from the cerebrum, and look down. It will appear as four rounded lobes with the two anterior lobes being larger than the two posterior ones. The four lobes are called corpora quadrigemina.

b. From the ventral view—Start at the anterior end of the brain, and find the olfactory bulbs. They represent Cranial Nerve I, and they convey the sense of olfaction from the nose to the brain. The optic chiasm represents Cranial Nerve II, and conveys the sense of sight from the retina to the occipital lobe of the brain. The word "chiasm" means cross, and the optic nerve tracts cross over at the level of the optic chiasm. This means that a portion of the visual field in one eye is perceived by the opposite side of the brain. Look directly posterior to the optic chiasm, and you should see the pituitary gland. The pons is a rounded section of the brainstem that functions to transfer information between the cerebellum, the cerebrum, and the medulla oblongata. It resembles a bridge in appearance. The following can also be seen from the ventral view: cerebrum, cerebellum, midbrain, medulla, temporal lobe, and lateral fissure.

c. From the sagittal view—A sagittal section is a cut in a longitudinal plane of space. Typically in brain sections, the sagittal section is a midsagittal section, which means that it divides the cerebral hemispheres and cerebellum into equal right and left halves. The corpus callosum can be seen from this view. It is made of white matter, and looks like an elongated "C," which has been rotated and turned upside down. It functions to connect the two cerebral hemispheres, and will be located at the base of the cerebral cortex. The thalamus is found below the corpus callosum, and appears as an oval-shaped structure. It forms the roof of the third ventricle, which is a cavity of the brain filled with cerebrospinal fluid. The thalamus has a nickname, called the "Grand Central Station" of the brain, because it functions to receive all incoming sensory information. The information passes through the thalamus first, before being routed to other areas of the brain for perception. The hypothalamus is located under the thalamus, and makes up the floor of the third ventricle. The feelings of sexual desire and rage are located in this part of the brain. In addition, the centers of thirst and satiety are found here. The fourth ventricle is a small cavity between the cerebellum and the medulla. It connects to the third ventricle via the cerebral aqueduct. In the sagittal section, you should also be able to locate the cerebrum, cerebellum, medulla, pons, midbrain, and the olfactory bulb.

The function of each structure to be labeled appears in the following table.

Through experimentation and study of brain injuries, many of the functional brain areas have been located. Figure 13.7 contains stippling, indicating some of the functional areas. Label each area with its function. All areas not stippled represent association areas, and are made up of neurons contributing to higher intellectual functions such as memory, reasoning, learning, and imagination.

TABLE 13.1 Structure and Function of Human Brain Regions

Structure	Function
Frontal lobe	Conscious thought; association area; motor area at posterior end.
Medulla Oblongata	Reflex center; controls heart rate, breathing, swallowing.
Corpus Callosum	Nerve fiber tracts connecting cerebral hemispheres.
Hypothalamus	Controls secretion of pituitary gland; appetite, water balance, blood pressure and body temperature, sleep rhythm.
Thalamus	Integrating center for sensory impulses.
Cerebellum	Coordination of motor movements and equilibrium.
Occipital lobe	Visual receiving center.
Ventricles	Cavities inside brain, contain cerebrospinal fluid (CSF).
Olfactory bulb	Sense of smell, Cranial Nerve I.
Central fissure	Separates frontal lobe from parietal lobe; separates motor and sensory areas of the brain.
Temporal lobe	Hearing (upper area) and smell (medial portion).
Pituitary gland	Master endocrine gland, secreting numerous hormones.
Parietal lobe	Contains sensory areas of brain in anterior portion; remainder is an association area.
Midbrain	Reflex center for hearing and vision.
Optic Chiasm	Optic nerve tracts cross; right tract leads to left occipital lobe, and v.v.
Pons	"Bridge"; connects cerebrum, cerebellum, and medulla.

Name: _____

QUESTIONS

The simplest type of reaction in the human nervous system is the **reflex**, which is an involuntary response that is always the same for a given stimulus. The simplest type of reflex involves a sensory and a motor neuron. Postural reflexes are of the two-neuron type. Most reflexes transmitted through the spinal cord involve at least three types of neurons: sensory, motor, and association. A classic example of a reflex is the withdrawal reflex of one's hand upon contact with a painful stimulus.

1. How is it possible for the hand to withdraw automatically from a stimulus before the sensation of pain is felt?

2. Trace the nerve pathway through which the stimulus must travel in order to bring about hand movement (see previous Figure 13.3). Start with the receptors.
   a. _____
   b. _____
   c. _____
   d. _____
   e. _____

3. What part of the CNS is directly involved in a simple reflex involving the hand?

4. When the pain from this stimulus has been consciously realized and identified, what portions of the CNS have been involved?

14 LABORATORY

The Nervous System: Sense Organs

A. Instructional Objectives

After completing this exercise, the student should be able to:

1. Identify from eye models, drawings, and preserved eyes the following structures, and know their function:
   a. Accessory eye structures—eyebrows, eyelashes, eyelids, conjunctiva, lacrimal apparatus (lacrimal gland), lacrimal ducts, lacrimal sac, and nasolacrimal duct
   b. External eye structures—sclera, cornea, optic nerve, eye muscles
   c. Internal eye structures—aqueous humor, suspensory ligaments, ciliary muscles, vitreous humor, sclera, choroid, retina, blind spot, fovea centralis, rods, cones, lens, iris

2. Explain the following about the eye:
   a. Accommodation
   b. Astigmatism
   c. Blind spot

3. Identify from models and drawings of the ear, and know the function of:
   a. Outer or external ear—pinna (auricle), auditory canal, tympanic membrane
   b. Middle ear—malleus, incus, stapes, oval window, Eustachian tube
   c. Inner ear—labyrinth, cochlea, organ of Corti, auditory nerve, vestibular nerve, cochlear nerve, round window, semicircular canals, saccule, utricle

4. Define all terms in the exercise, plus the following:

Biconvex disc	"white of the eye"	tapetum
Snellen eye chart	c.p.s. (cycles per second)	basilar membrane
Otoliths	auditory ossicles	perception
Near point		

B. Introduction

The sense organs are those that allow us to see, hear, taste, touch, and smell. The word "sense" or "sensation" is defined as a feeling: the translation into consciousness of the effects of a stimulus exciting any of the organs of sense. Perception is the mental process of becoming aware of, or recognizing, a sensation or an object. Perception is possible because of afferent nerve pathways (nerves) between the sense organs and the brain. When a sense organ is stimulated, this initiates a nerve impulse at the level of the receptor, and the impulse travels toward the central nervous system. Once it reaches the brain, it is analyzed and interpreted, and is perceived.

1. Per 2 students: 1 mammalian (sheep) eye, and dissecting tray
2. Per laboratory: 1 beaker, eye models, eye charts, ear models, ear charts

Methods

Obtain a preserved eye from the eye bucket. Study the external eye structures before removing the fat and muscles. Then, cut off excess fat and muscles around the eye before attempting internal dissection. There are two figures for study: Figure 14.1 and Figure 14.2. Study Figure 14.1, and label Figure 14.2 as you dissect. You will find this figure in a later section of this exercise.

1. **Accessory structures of the eye: Eyebrows** and **eyelashes** provide a moderate amount of protection against the entrance of foreign particles, such as dirt, dust and excess sweat. **Eyelids** are made of skin and striated muscle. They are covered on their deep surfaces by **conjunctiva**, which is a mucous membrane that is folded over the eyeball. This serves to protect the anterior portion of the eye. The lacrimal apparatus consists of a lacrimal gland, lacrimal sac, **lacrimal ducts, and a nasolacrimal duct.** *Lacrimal* means tears; they work together to form tears to bathe the eyeball and keep it moist. The excess fluid drains into the nasal passage.

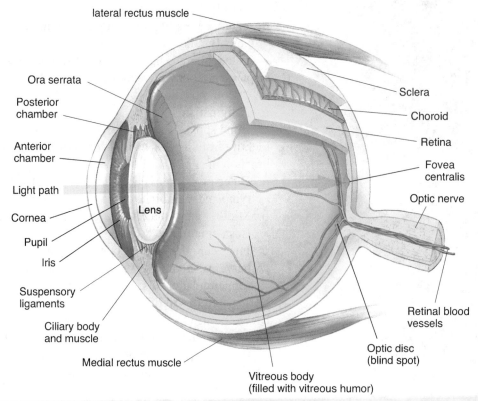

FIGURE 14.1   Eye Anatomy
Copyright © Kendall/Hunt Publishing Company.

2. **External eye structure**:
   a. Before cutting into the eyeball, observe the tough outer white membrane, called the sclera. It protects the inner structures, and helps maintain rigidity of the eyeball.
   b. The sclera is transparent toward the front, or anterior, end of the eyeball. The transparent area is called the cornea. A common eye defect called astigmatism is due to an irregularity in the curvature of the cornea.
   c. The pupil is seen from the front as a dark spot in the center of the eye. It is an opening in the center of the iris, through which light enters the eye.
   d. The iris is a thin, circular, pigmented diaphragm. It controls the amount of light admitted to the eye. The pigment of the iris accounts for eye color.
   e. Try to locate six muscles branching outward from the eyeball. These ocular muscles account for eyeball movement from side to side and up and down.
   f. Note the optic nerve (Cranial Nerve II), which is a large tough cord that exits from the back of the eyeball. The nerve extends from the retina to the occipital lobe of the brain, and transmits sensory impulses.

3. **Internal eye structure**:

   Take a scalpel, and divide the eye into two equal right and left halves, by cutting through the sclera. Cut from the optic nerve forward through the center of the pupil. The space anterior to the lens is filled with a watery fluid, called aqueous humor, and will probably spill out as the dissection continues. The crystalline lens lies in back of the iris and pupil. It is a biconcave disc, and will be hard to the touch. It is transparent, but may be cloudy because of preservative. It is made of concentric layers of protein, and is held in place by the suspensory ligament and ciliary muscles. These muscles are responsible for changing the shape of the lens for focusing at different distances. The lens is normally elastic, and can adjust to form different shapes, due to the action (contraction and relaxation) of the ciliary muscles. A jelly-like substance called vitreous humor fills the space between the lens and the back of the eyeball.

   The wall of the eyeball consists of three layers. The outer layer (the sclera) has already been examined. This is the "white of the eye," and is the visible portion of the sclera around the cornea. The middle layer of the eye is opaque, a dark black or blue color, and is called the choroid layer. Its purpose is to absorb light and keep it from scattering. It is richly supplied with blood vessels. In mammals, the tapetum is a rich iridescent area, and helps them to see better at night. Its color is similar to mother-of-pearl.

   The inner layer is the retina, which is actually nervous tissue, as it is an extension of the optic nerve. It contains several layers: a pigmented layer that adheres to the choroids, and a layer that consists of visual receptor neurons (rods and cones), bipolar neurons, and neural ganglia that are directly attached to fibers of the optic nerve. The retina is fragile, and when the eye is dissected and the vitreous humor detaches, the retina will crumple and fall away from the back of the eye, similar to wrinkled tissue paper. The continuity of the retina's surface is interrupted by two structures, the optic nerve, or blind spot, and the fovea centralis. The specific area on the retina where the optic nerve exits is called the blind spot because it contains no rods or cones. Thus, no receptors for dim and bright light exist in this area. The fovea, in contrast, contains all cones, and represents that part of the eye where critical, sharp, colored vision occurs.

4. **Compare the preserved eye with the eye model, and label Figure 14.2.**
5. **Function of the mammalian eye:**

   As two students begin the following parts of this exercise, study Figure 14.3, "Focusing—Close and Distant."
   a. Accommodation—The eye is capable of focusing on near and distant objects. The focus is possible through accommodation, which involves a change in the curvature of the lens. The lens can stretch and relax because of the ligament and muscle control. To observe and practice:
      1. Face an object approximately 20 feet away, and hold your hand with outspread fingers at a distance of half a foot from your face. Close your eyes. Immediately upon opening them, do you see the fingers distinctly, or do you see the distant object?

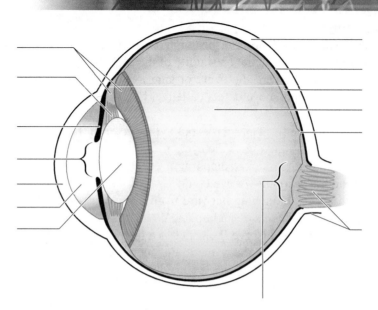

FIGURE 14.2   Structures of Eye
Copyright © Kendall/Hunt Publishing Company.

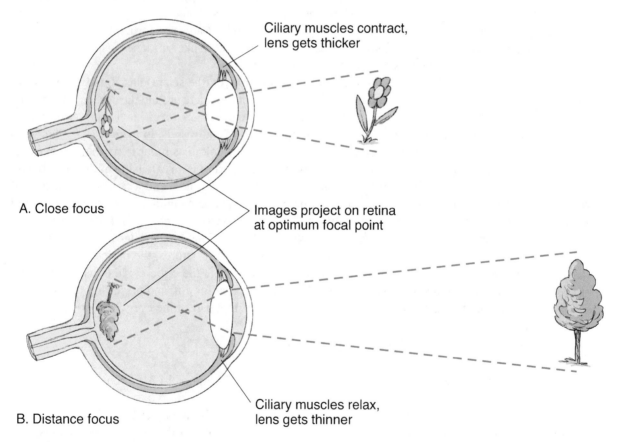

FIGURE 14.3   Focusing—Close and Distant
Copyright © Kendall/Hunt Publishing Company.

2. Look again at the distant object, and after four or five seconds, look at the fingers. Do you notice any difference in the "feeling" experienced in changing from the far to the near object? Was there a lapse of time? What does this indicate?

3. In moderate light, one student (the subject) should look at an object at least 20 feet away. Another student should examine the size of the subject's pupils. The subject should then look at a pencil held about 10 inches from the face. The second student should note the size of the pupils of the subject. What changes occur?

b. Astigmatism—This is the name for a condition in which the refracting surface of the cornea or lens is not a perfect sphere, but is curved irregularly.

   1. Observe an astigmatism chart in the lab at a distance of approximately 20 feet, using one eye.

   2. Notice if some of the radiating lines appear darker, or more distant than others. Such an appearance is indicative of astigmatism. Test the other eye in the same manner.

   3. If you do not have an astigmatism, place a small beaker of water before your eye, and repeat the experiment. The refraction of light through the water will simulate an astigmatism.

c. Blind spot—the small area, insensitive to light, at the back of the retina where the optic nerve exits.

   1. With the left eye closed, and the laboratory book held about 20 inches from the face, look at the cross shown in Figure 14.4.

FIGURE 14.4   Demonstration of blind spot

   2. Gradually bring the book closer to you as you keep your eye fixed upon the cross, until the dot disappears.

   3. Bring the page closer still. What happens?

d. Demonstration of the Snellen Eye Chart: The Snellen Eye Chart, named after Hermann Snellen who developed it in 1862, is an eye chart that is used to measure visual acuity. It contains 11 lines of block letters, formally called "optotypes." The standard procedure is for the patient to stand 20 feet away from the chart, cover one eye at a time, and read the letters in each row. The patient starts at the top and continues reading each subsequent row, with decreasing letter sizes. The smallest row that can be read accurately is indicative of the patient's visual acuity in that eye. The largest letter on the chart usually represents a visual acuity of 20/200. A patient who is "legally blind" is one who cannot read the top letter, even with corrected lenses. Students work in pairs, and observe the Snellen chart hanging on the wall in the lab. The chart can also be studied in Figure 14.5.

6. Structure of the ear—The ear is the organ of hearing and equilibrium. Its sensory components are buried deep within the bones of the skull, and many accessory structures are necessary to transmit sound waves (waves of compressed air) from the external environment to the sensory cells. The human ear can detect vibrations (or hear) in the range between 20 and 20,000 cycles (vibrations) per second. Middle C corresponds to 256 c.p.s. Study a model of the human ear in the lab; study Figure 14.6; then locate the parts appearing in boldface in the following discussion. Label Figure 14.7, found in a later section of the exercise.

a. Outer ear:

   1. The visible part of the outer ear is a skin-covered, cartilaginous flap called the pinna. It is modified to direct and concentrate sound waves into the inner parts of the ear.

   2. The auditory canal leads from the pinna to the middle ear, and serves as a passageway for the sound waves. In a living specimen, there are numerous glands found in the skin surface of the canal, which secrete a wax that lubricates the canal and protects it. The wax will trap debris to keep it from traveling toward the ear drum, or tympanic membrane.

Visual Acuity Chart - Approximate Snellen Scale
For Educational Purposes Only

$\dfrac{20}{200}$ **E** $\dfrac{200 \text{ ft}}{61 \text{ m}}$

$\dfrac{20}{100}$ **H N** $\dfrac{100 \text{ ft}}{30.5}$

$\dfrac{20}{70}$ **D F N** $\dfrac{70 \text{ ft}}{21.3 \text{ m}}$

$\dfrac{20}{50}$ **P T X Z** $\dfrac{50 \text{ ft}}{15.2 \text{ m}}$

$\dfrac{20}{40}$ **U Z D T F** $\dfrac{40 \text{ ft}}{12.2 \text{ m}}$

$\dfrac{20}{30}$ **D F N P T H** $\dfrac{30 \text{ ft}}{9.1 \text{ m}}$

$\dfrac{20}{20}$ **P H U N T D Z** $\dfrac{20 \text{ ft}}{6.1 \text{ m}}$

$\dfrac{20}{15}$ **N P X T Z F H** $\dfrac{15 \text{ ft}}{4.6 \text{ m}}$

FIGURE 14.5   The Snellen Eye Chart

3. Note the tympanic membrane, a very thin membrane that starts vibrating when sound waves strike it. The vibrations are transmitted via the ear bones to the inner ear. When the strength of the vibrations increase, the pitch becomes higher.

4. The eardrum is stretched across a cartilaginous ring, and is at the junction of the auditory canal and the middle ear.

b. **Middle ear:**

1. The middle ear is a small chamber containing three tiny bones connected in a series, the **malleus (hammer), incus (anvil),** and the **stapes (stirrup).** They transmit the mechanical vibrations of the eardrum across the middle ear cavity into the inner ear. Locate these three bones, and note that the malleus is in contact with the tympanic membrane and the incus. Note that the stapes is in contact with the incus and the opening called the **oval window** into the inner ear.

2. Observe the **Eustachian tube,** which connects the middle ear to the pharynx (throat). This tube equalizes pressure on the two sides of the eardrum. If the middle ear were completely closed, small changes in atmospheric pressure would cause pronounced and painful bulging, or caving in, of the eardrum.

c. **Inner ear**—The inner ear consists of a complicated group of interconnected canals and sacs, often referred to as the **labyrinth.**

1. The part of the labyrinth concerned with hearing is a spiral-shaped coiled tube of two and a half turns, resembling a snail's shell, called the **cochlea.** Remove the parts of the inner ear on the model to locate this structure. The cochlea contains three fluid-filled chambers, the **scala tympani,** the **scala vestibule,** and the **cochlear duct.** The cochlear duct is separated from the scala tympani by the basilar membrane. The membrane contains the receptors for hearing called the **organ of Corti.** The receptors translate sound waves into nerve impulses.

2. Note the **auditory nerve** (Cranial Nerve VIII), also known as the vestibulocochlear nerve. It is associated with the cochlea, and contains another nerve called the **vestibular nerve.** The vestibular nerve arises in the semicircular canals and the vestibule, and the cochlear nerve arises from the cochlea. Thus, the presence of the two nerves means that the auditory nerve conveys the senses of hearing and balance from the inner ear to the brain.

3. Note the **round window** on the cochlea. Its function is to help maintain pressure equilibrium within the cochlea.

4. The part of the labyrinth concerned with equilibrium, or balance, consists of **three semicircular canals,** and two small sacs, called the **saccule** and the **utricle.** The canals are arranged so that each is at right angles to the other two. They contain a fluid called **endolymph** and clumps of hairlike cells. These cells are stimulated by head movements that cause the fluid to move in response to the turning of the head. A movement of the head in any direction will stimulate the movement of the fluid in at least one of the canals. These hairlike cells are connected to the sensory nerve fibers and send impulses to the brain, where the head movements are perceived. The utricle and saccule are hollow sacs lined with sensitive hairlike cells. These small structures also contain **otoliths** (ear stones made of calcium carbonate). Normally, the pull of gravity causes the otoliths to press against the sensory cells, and they will move freely as the head and body turn in different directions. Label Figure 14.7.

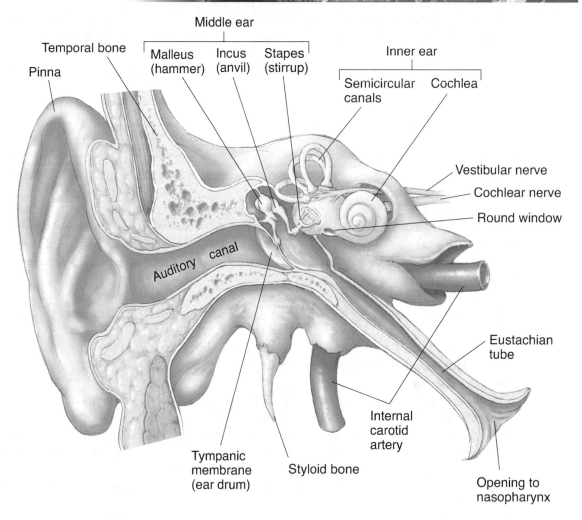

FIGURE 14.6   Ear Anatomy
Copyright © Kendall/Hunt Publishing Company.

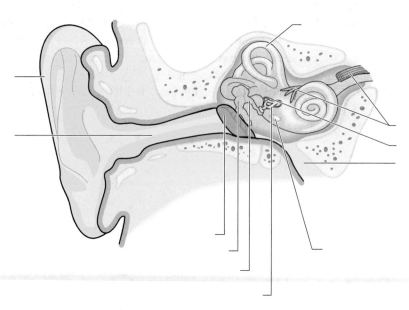

FIGURE 14.7   Structures of Human Ear

Name: _____

QUESTIONS

1. A clouded lens is a cataract, and results from the breakdown of protein in the lens in old age, overexposure to bright light, trauma, diabetes, and so forth. To treat "cataracts," what portion of the eye must be removed?

2. How could a person who had a cataract operation be able to focus on an image on the retina of the eye?

3. The closest distance at which an object appears to be in sharp focus is called the near point. For example, at 20 years of age, the normal near point is 3½ inches.
   At 30 years of age, the normal near point is 4½ inches.
   At 40 years of age, the normal near point is 6¾ inches.
   At 50 years of age, the normal near point is 20¼ inches.
   At 60 years of age, the normal near point is approximately 33 inches.

   How could one account for these changes?

   What is your near point?
   Does it fall within the figures listed above?

4. List in sequence the structures through which sound waves and nerve impulses pass from the moment a word is spoken to the perception of the word.

5. Man has become used to movement in the horizontal plane, which stimulates certain semicircular canals, but he is unused to vertical movements parallel to the long axis of the body. Explain why some persons will become nauseated from the rocking of a boat, an elevator ride, or a fast ride on a merry-go-round.

PSALM
50 15

15 LABORATORY

Mitosis

A. Instructional Objectives: After completing this exercise the student should be able to:
   1. Distinguish between mitosis and cytokinesis.
   2. Identify the stages of animal cell mitosis from slides and models.
   3. Identify the structures involved in mitosis.
   4. Distinguish between animal and plant mitosis.

B. Materials

   Models of plant cell mitosis   Models of animal cell mitosis

   Prepared slides of onion   Prepared slides of whitefish
   root tip mitosis   blastula mitosis

C. Introduction

   All eukaryotic multicellular organisms (plants, animals, fungi, and protistans) use mitosis for growth, repair, and/or asexual reproduction. The value of asexual reproduction is that it is fast and little energy is wasted. A large number of offspring can be produced in a relatively short time. The disadvantage of asexual reproduction is that the daughter cells are genetically identical to each other and to the mother cell. Being genetically identical in an ever-changing environment has a significant disadvantage. Since all of the cells are identical and have the same genetic material, an environmental challenge could be detrimental to all organisms.

   Mitosis refers to the division of the nucleus, whereas the *division of the cytoplasm* refers to cytokinesis. In mitosis, two identical daughter cells are formed. Each daughter cell has the same number and kind chromosomes as the mother cell. The daughter cells are clones of each other. Refer to Table 15.1 and Figure 15.1.

   Mitosis consists of five different recognizable phases (six if daughter cells are counted as a phase). These phases are part of a continuous process. Prior to the slide being prepared, mitosis has been halted. When you view a slide under the

TABLE 15.1   Characteristics of Mitosis

Number of divisions	Number of cells formed	Number of chromosomes
1	2	Remains constant

microscope, you are seeing the tissue as if you were viewing a photograph in stop-action. The characteristics of each phase in animal and plant cells are listed in Table 15.2.

D. Plant cell division

Plant cell chromosomes are usually larger than animal cell chromosomes. This is the reason why we will start with the prepared and stained slide of *Allium* (onion) root tip. The root tip of plants is an area of rapid cell division and growth due to cells rapidly dividing. It is called an apical meristem. The end of the root is covered by a root cap. These cells are relatively thin and stratified to protect the meristem. Most cells are in interphase, so you will need to scan the tissue to find all of the stages of mitosis. Remember that you need to be able to recognize the phase of mitosis under the microscope. Remember also that plant cells do not have centrioles and asters are also not present in plant cells; plant cells have cell walls. In cytokinesis the cell plate forms in the middle of the dividing cell in late telophase. As the cell plate forms, it grows toward the cell wall.

E. Animal cell division

Select a slide of the whitefish blastula. The blastula is an early stage of embryonic development where the cells are rapidly dividing. The cells contain oil droplets, which are a source of energy in the cytoplasm. These oil droplets have a purple stain. Since the cells are rapidly dividing, it might be difficult to find a cell in interphase. The nucleus is not visible in this stage of mitosis. The animal cell chromosomes are smaller than those of the plant. In this tissue it is very difficult to pick an individual chromosome. Remember that animal cells have centrioles. The incomplete spindle fibers that radiate from the centriole toward the cell membrane are called asters. Also remember that in cytokinesis the cell membrane pinches off like a purse string.

1. Draw and label the plant and animal cells representing the phases of mitosis.

   Interphase: chromatin, nucleolus (if visible), nuclear envelope, centrioles (if present)

   Prophase: chromosomes, nucleolus (if visible), nuclear envelope, centrioles (if present)

   Metaphase: equatorial plane, spindle fibers, asters (if present)

   Anaphase: spindle fibers, asters, daughter chromosomes

   Cytokinesis: cell plate (plants), cleavage furrow (animals)

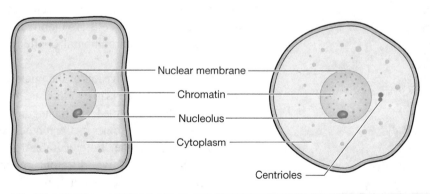

Plant   **Animal**

FIGURE 15.1   Interphase

TABLE 15.2   Phases of Mitosis

Phase	Animal	Plant
Interphase	1. The nucleus is distinct. 2. Chromatin is present. 3. Chromosomes are replicated. 4. Nucleoli are visible. 5. Centrioles are visible next to the nucleus.	1. The nucleus is distinct. 2. Chromatin present. 3. Chromosomes are replicated. 4. Nucleoli are visible.
Prophase	1. Chromatin coils to form visible rod-shaped chromosomes. 2. Nucleoli disappear. 3. Nuclear membrane disappears. 4. Centrioles with *asters* (incomplete spindle fibers) divide and migrate to opposite poles of the cell.	1. Chromosomes appear. 2. Nucleoli disappear. 3. Nuclear membrane disappears.
Metaphase	1. Spindle fibers radiate from the centrioles. 2. Chromosomes align on the equatorial plate. 3. Spindle fibers attach to the *centromeres*.	1. Spindle fibers form. 2. Chromosomes align on the equatorial plate. 3. Spindle fibers attach to the *centromeres*.
Anaphase	1. *Centromeres* split and separate. 2. Daughter chromosomes move by the aid of the spindle fibers to opposite poles of the cell.	1. *Centromeres* split and separate. 2. Daughter chromosomes move by the aid of the spindle fibers to opposite poles of the cell.
Telophase	1. Daughter chromosomes reach opposite poles. 2. Chromosomes uncoil. 3. Nuclear membrane re-forms. 4. Nucleoli become visible.	1. Daughter chromosomes reach opposite poles. 2. Chromosomes uncoil. 3. Nuclear membrane re-forms. 4. Nucleoli become visible.
Daughter Cells	1. Identical cells that are half the size of the mother cell. 2. Identical number of chromosomes as the mother cell had. 3. Chromosomes are identical to the mother cell's chromosomes.	1. Identical cells that are half the size of the mother cell. 2. Identical number of chromosomes as the mother cell had. 3. Chromosomes are identical to the mother cell's chromosomes.

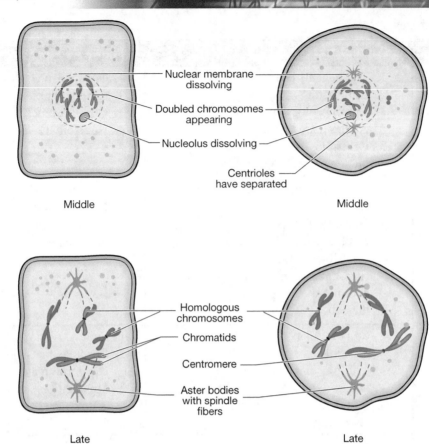

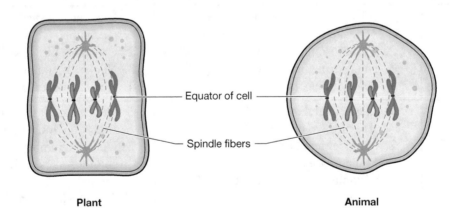

Plant   **Animal**

FIGURE 15.2   Prophase
Copyright © Kendall/Hunt Publishing Company.

Plant   **Animal**

FIGURE 15.3   Metaphase
Copyright © Kendall/Hunt Publishing Company.

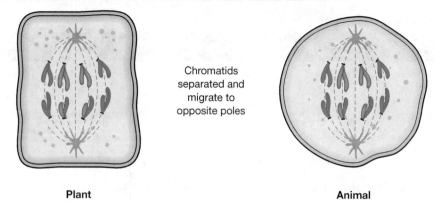

Chromatids separated and migrate to opposite poles

Plant

Animal

FIGURE 15.4   Anaphase
Copyright © Kendall/Hunt Publishing Company.

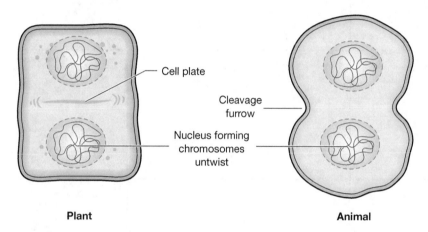

Cell plate

Cleavage furrow

Nucleus forming chromosomes untwist

Plant

Animal

FIGURE 15.5   Telophase
Copyright © Kendall/Hunt Publishing Company.

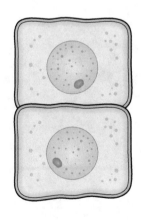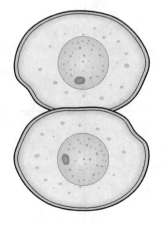

Plant

Animal

FIGURE 15.6   Daughter Cells
Copyright © Kendall/Hunt Publishing Company.

TABLE 15.3 Plant and Animal Stages

Phase	Plant Cell Drawing	Animal Cell Drawing
Interphase		
Prophase		
Metaphase		
Anaphase		
Telophase		
Daughter Cells		

Name: *Christian Sweeney*

CONLUSIONS AND QUESTIONS

1. What is the importance of chromosome duplication prior to mitosis?

2. How does the genetic material contained within the chromosomes compare between the mother cell and the daughter cells?

3. Distinguish between the following terms:
   a. chromatin
   b. chromosome
   c. chromatid

4. List three differences found in the mitosis of plants and animals.
   a. _____
   b. _____
   c. _____

5. Write the correct phase of mitosis for each event.
   _____ a.   DNA and chromosome replication occurs.
   _____ b.   Chromatin is present.
   _____ c.   Chromatin begins to condense
                      into chromosomes.
   _____ d.   Nucleus begins to disappear.
   _____ e.   Chromosomes appear.
   _____ f.   Chromosomes begin to migrate to the equatorial plane.
   _____ g.   Chromosomes align on the equatorial plane.
   _____ h.   Centromeres split lengthwise and daughter chromosomes separate.
   _____ i.   Daughter chromosomes migrate to opposite ends of the cell.
   _____ j.   Chromosomes uncoil and nucleus re-forms.
   _____ k.   Cleavage furrow in animal cells forms. Cell plate in plant cells forms.
   _____ l.   Two identical cells formed.

16 LABORATORY

Meiosis

A. Instructional Objectives: After completing this exercise the student should be able to:

1. Understand the overall process of meiosis by being able to explain chromosomal behavior during the stages of meiosis I and II.
2. Label the cells of each meiosis phase as haploid or diploid.
3. Identify cells in the following stages: prophase I, metaphase I, anaphase I, telophase I, prophase II, metaphase II, anaphase II, telophase II, and daughter cells.
4. Use the following terms: parent cell, homologous chromosomes, synapse, tetrad, dyad, and gamete.
5. Follow the movement of two pairs of homologous chromosomes through the meiotic process using the meiosis demonstration kit.
6. Diagram mitosis and meiosis I and II using one pair of homologous chromosomes.
7. Recognize and understand the differences in the two types of cell divisions.
8. Define and understand the following terms: gonads, sperm, eggs, meiosis, somatic cell, diploid (2n), haploid (n), homologous pair of chromosomes, replication, and fertilization.

B. Introduction

All organisms, including animals, plants, fungi, and protist, must grow and replace lost or damaged cells. They do this by the process of mitosis. However, when an organism wants to reproduce itself, it may do so through asexual reproduction or sexual reproduction. During sexual reproduction, a cell from the mother and a cell from the father must come together and fuse so that the chromosomes from each parent equally contribute to the genetic makeup of the offspring. If the cells that fuse during fertilization came directly from the parents' body (somatic) cells, then fertilization would result in an offspring whose cells each had double the number of chromosomes. This type of offspring could not survive.

Therefore, organisms have developed a means of producing special sex cells called gametes. Gametes are produced when the gonads of the mother and of the father each make special sex cells through meiosis, a special type of cell division. During meiosis, the number of chromosomes is halved. A "parent cell" undergoes two cell divisions to produce four daughter cells instead of two daughter cells like in mitosis. The male gamete is called the sperm, and the female gamete is called the egg.

Somatic (body) cells of humans contain the normal number of chromosomes, which is 46. This is known as diploid, or $2n$, because half of the set of chromosomes came from the mother, and the other half came from the father. A cell with half the normal number of chromosomes, which is 23, is known as haploid, or n. The human egg and sperm each have only 23 chromosomes and are therefore haploid cells.

When the egg and sperm fuse during sexual reproduction, the normal number of chromosomes is restored and the resulting offspring is now diploid with 46 chromosomes. Figure 16.1 illustrates this series of events; the numbers represent the chromosome complement per cell, where $2n = 46$ and $n = 23$.

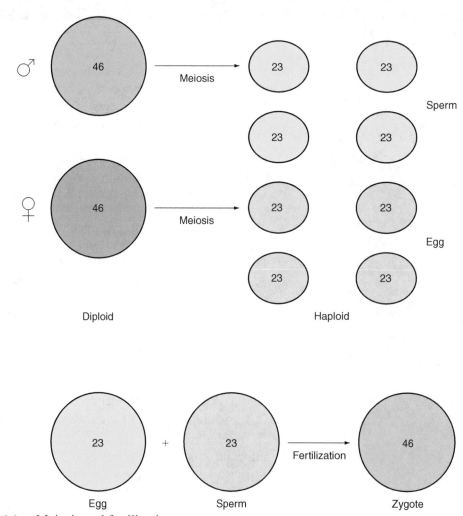

FIGURE 16.1.   Meiosis and fertilization

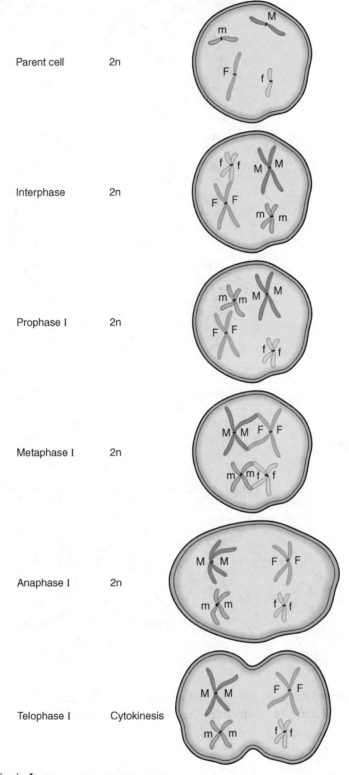

FIGURE 16.2A Meiosis I. © *Kendall/Hunt Publishing Company.*

Prophase II   n

Metaphase II   n

Anaphase II   n

Telophase II   Cytokinesis

Daughter cells   n

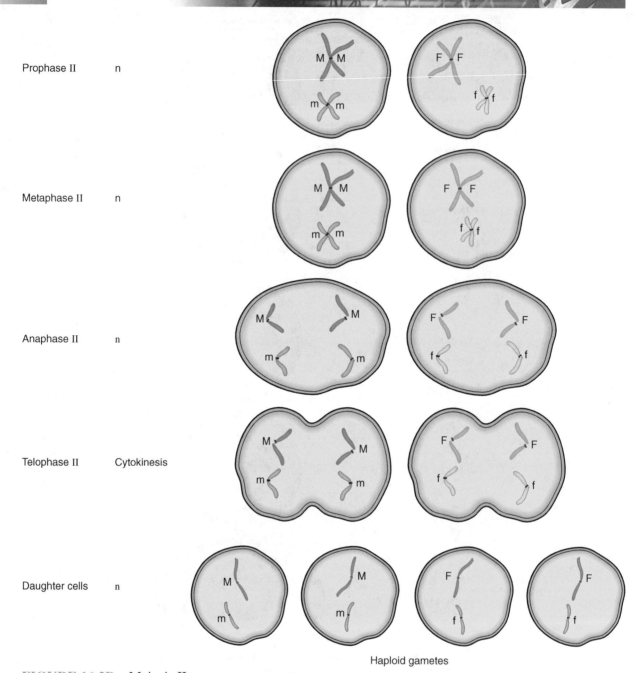

Haploid gametes

FIGURE 16.2B   Meiosis II. © *Kendall/Hunt Publishing Company.*

The process of meiosis is divided into two major steps: Meiosis I and Meiosis II. Meiosis I and II are each divided into phases. Figure 16.2 illustrates the chromosomal behavior in each of the phases.

Interphase

Brief description: During interphase, cells are not actively dividing but are carrying on normal metabolic activities.

1. Nucleus is present and obvious. DNA is referred to as chromatin, which will later become chromosomes when the cell gets ready to divide.
2. Nuclear membrane, also known as the nuclear envelope, is intact and visible.
3. DNA duplicates itself.
4. Centrioles are present in animal cells. Plant cells do not have centrioles.

Prophase I

Brief description: During prophase, the cell modifies its DNA into chromosomes and prepares for cell division.

1. Chromatin condenses to become plainly visible chromosomes. Each chromosome is made of two sister chromatids held together by a centromere. Chromosomes are found in pairs called homologous chromosomes. One chromosome came from the mother; the other came from the father. Homologous chromosomes have genes for the same traits.
2. Homologous chromosomes pair up (synapse). Each group of four chromatids is called a tetrad.
3. Nuclear membrane disintegrates.
4. Nucleoli fade away.
5. Centrioles of animal cells move toward opposite poles of the cells and form aster bodies.
6. Spindle fibers appear and begin to attach to the centromeres of the chromosomes.

Metaphase I

1. Homologous chromosome pairs line up next to each other at the midline of the cell.
2. Spindle fibers finish attaching to the centromeres of each chromosome. They will later help guide the chromosomes toward opposite poles of the cell during cell division.

Anaphase I

Homologous chromosomes separate and begin to move toward opposite poles of the cell.

Telophase I and Cytokinesis

1. Homologous pairs are separated and are at opposite ends of the cell.
2. In the animal cell, cytoplasm pinches in to begin separation into two cells (cleavage furrow).
3. In the plant cell, the cell plate forms between two new cells.

Prophase II

1. Chromatin condenses to become plainly visible chromosomes. Each chromosome is made of two sister chromatids held together by a centromere.
2. Nucleoli fade away.
3. Centrioles of animal cells move toward opposite poles of the cells and form aster bodies.
4. Spindle fibers appear and begin to attach to the centromeres of the chromosomes.

Metaphase II

1. Double-stranded (dyads) chromosomes line up end to end along the midline of the cell.
2. Spindle fibers finish attaching to the centromeres of each chromosome. They will later help guide the chromosomes toward opposite poles of the cell during cell division.

Anaphase II

Sister chromatids separate at the centromere and move toward opposite ends of the cell (centromeres separate lengthwise).

Telophase II and Cytokinesis

1. Chromatids have moved to opposite ends and are now called chromosomes.
2. Cleavage furrow forms.
3. Cell plate forms.

Daughter Cells

Four haploid gametes result.

C. Materials
   1. Meiosis demonstration kits
   2. Cell outline sheets for meiosis and fertilization

D. Methods

   Before using the chromosome demonstration kits, be certain that you understand and have a good idea of chromosomal movement during the process of meiosis. Use Figure 16.2 to help you. For simplicity, only two homologous pairs of chromosomes will be used in the demonstration kit. The long chromosomes represent one pair; the short chromosomes represent the second pair.

   1. Take all the chromosomes and four centrioles out of the bag. You should have eight single-stranded or four double-stranded chromosomes. The long ones should both be the same color; the short ones should both be the same color. Take two long single-stranded chromosomes and two short single-stranded chromosomes and assemble them in the "nucleus" of your cell. This represents the parent cell.

   2. Next, replicate the chromosomes. Take out two more long single-stranded chromosomes and pair one with each of the already existing chromosomes. Do the same with the short chromosomes. The chromosomes are now replicated and double-stranded. Each duplicated chromosome is made of two sister chromatids. This represents interphase.

   3. Now move the chromosomes so that the long duplicated chromosomes lie next to each other, lengthwise. Do the same with the shorter ones. This pairing of chromosomes is called a synapse. Place the pairs along the midline of the cell. This represents metaphase I.

   4. Begin moving the homologous pairs toward opposite ends of the cell. This represents anaphase I and telophase I.

   5. Now move the separated double-stranded chromosomes into their two separate cells. Each cell should have one long duplicated chromosome and one short duplicated chromosome.

   6. Each of the two cells immediately undergoes a second division. Line the chromosomes up end to end along the midline of the cell. This represents metaphase II.

   7. Separate the sister chromatids at the centromere and begin moving the resulting chromosomes toward opposite ends of the cells. This represents anaphase II.

   8. Finally, place each set of chromosomes in one of the four cells. All daughter cells should have chromosomes of the same number. They are haploid. Each cell should have one long and one short chromosome. This represents telophase and cytokinesis.

CHART 16.1   Comparison/contrast Mitosis and Meiosis

Meiosis	Mitosis
Formation of gametes	Growth and repair
2 divisions (Meiosis I and II)	1 division
Parent not identical to daughter cells	Parent identical to daughter cells
4 daughter cells	2 daughter cells
Parents are diploid (*2n*)	Parents are diploid (*2n*)
Daughter cells are haploid (*n*)	Daughter cells are diploid (*2n*)
Crossing over during prophase I	No crossing over

Name: _____

QUESTIONS

1. List four ways in which meiosis differs from mitosis.

2. Why is it important to reduce the chromosome number from diploid to haploid when producing gametes?

17

LABORATORY

Reproduction in Humans

A. Instructional Objectives

After completing this exercise, the student should be able to

1. Identify the following male reproductive structures from models, slides, or drawings and know their functions:
   a. Microscopic view of the testis: seminiferous tubules, spermatogenesis, spermatids, interstitial cells.
   b. Parts of sperm: head, acrosome, midpiece, flagella.
   c. Anatomy of the male reproductive organs: penis, scrotum, testis, epididymis, vas deferens, seminal vesicle, ejaculatory duct, prostate gland, Cowper's gland, urethra, corpus cavernosa, corpus spongiosum, prepuce, glans penis, bladder.

2. Identify the following female reproductive structures from models, slides, or drawings and know their functions:
   a. Microscopic view of the ovary: oogenesis, oocyte, primary follicle, developing follicle, Graafian follicle, ovulation, corpus luteum.
   b. Anatomy of the female reproductive organs: ovary, fimbriae, fallopian tube, uterus, perimetrium, myometrium, endometrium, cervix, vagina, clitorus, labium minora, labium majora.

3. Define the following terms: testosterone, circumcision, semen, hyaluronidase, sterile, Pap smear, menstruation, polar body, zygote, fertilization.

B. Introduction

With humans, as with all other organisms, the most important contribution we make to the species is the offspring we leave behind. These offspring will produce more offspring, who will in their turn reproduce, and assure the continuation of our species. Though each child is a precious and unique entity, every child is the result of a highly complex, efficient reproductive process common to us all. To better understand this process (which includes gamete formation, the union of these gametes, development, and birth), one should be familiar with the anatomy and physiology of the human male and female reproductive system.

C. Materials

Prepared slides: human sperm, mature testis, cat ovaries in various stages of development.
    Models of male reproductive system, female reproductive system, ovary section.

D. Methods

The male reproductive system starts in the paired testes (testis - singular; testes - plural) which produce the male gonads called sperm. The scrotum is the loose muscular sac outside the body that surrounds the testes and epididymis. The scrotum is divided in halves by a septum; each half contains a testis. Since the optimum temperature for sperm production is a few degrees lower than internal body temperature, the testes are suspended outside the body, facilitating sperm production. Fear of injury or severe cold temperatures will cause the cremaster muscle of the scrotum to contract and pull the testes closer to the body for protection and warmth.

The paired testes are slightly flattened oval glands surrounded by a fibrous connective tissue called the tunica albuginea. Inward extensions of the tunica form the septa, which divide each testis into 250 to 400 lobules. Within each lobule are one to four tightly packed, coiled, and twisted seminiferous tubules. Seminiferous tubules produce sperm, the male gametes, by the millions on a daily basis.

Spermatogenesis is the process of sperm production. Mitosis creates new normal body cells with DNA that is identical to the parent cell. Meiosis creates new gametes with half the amount of DNA of the parent cell. A baby gets half of its DNA from its father's sperm and half from its mother's egg.

Spermatogenesis takes place in the seminiferous tubules. Look at the seminiferous tubule or testis slide under the microscope. Locate one of the round tubules in cross section with the high powered lens. The outer layer of cells is spermatogonia; these cells divide by mitosis to form two cells. One daughter cell will remain a spermatogonium and continue dividing. The second daughter cell will become a primary spermatocyte. The primary spermatocyte divides (meiosis I) to become two secondary spermatocytes. The two secondary spermatocytes will divide again (meiosis II) to become four spermatids. Each spermatid contains the correct amount of DNA for the fertilization of an egg, but the spermatids are not yet motile or functional. Spermatids are located in the middle of the seminiferous tubule; you may or may not be able to distinguish their flagella. Every day a healthy adult male produces about 400 million sperm that take 64 to 72 days to fully mature.

Between the seminiferous tubules are clusters of cells called interstitial cells or Leydig cells. These cells produce testosterone, a male hormone responsible for the development of male reproductive organs and for the maintenance of male secondary sexual characteristics: facial hair, deep voice, heavy musculature, etc.

In the space below, draw and label a seminiferous tubule as seen under the microscope.

Look at a slide of human sperm using the high powered lens. The head of the sperm contains a nucleus filled with DNA and little else. Attached to the front end of the head is the acrosome, which contains enzymes that allow the sperm to penetrate and fertilize the egg. The midpiece of the sperm contains many mitochondria which produce ATP. The ATP is used as energy to propel the flagella, or tail.

In the space below, draw and label a sperm as seen under the microscope or on a model.

The seminiferous tubules of the testes drain into ducts, which drain into the epididymis. The epididymis is a comma shaped body, which is really one long coiled tube from 13 to 20 feet in length. Each epididymis starts at the top of a testis, extends downward along the back side of the testis, and then makes a hairpin turn upward. The epididymis secretes an alkaline fluid in which the sperm are suspended. Sperm must have an alkaline medium or they will remain immobile, even once they possess their flagella. Sperm become motile and are stored in the epididymis prior to ejaculation. The end of the epididymis becomes less coiled and its walls become thicker. At this point, the epididymis becomes the vas deferens.

The vas deferens (or ductus deferens) transports sperm from the epididymis to the urethra through peristaltic contractions of its smooth muscle. Each vas deferens extends upward from the epididymis into the interior of the body. They pass in front of and over the bladder, which stores urine. Each vas deferens turns downward behind the bladder where they are joined by the ducts which drain each of the two seminal vesicles.

The seminal vesicles are located on the posterior surface of the bladder, just in front of the rectum. The seminal vesicles secrete a viscous, alkaline fluid which contains fructose. The alkalinity of the fluid neutralizes the acidic conditions of the female vagina, while the fructose serves as a nutrient for the production of ATP for propulsion. If a man's secretions are not alkaline enough, he may be sterile because sperm are sluggish in acidic environments less than pH 6.0, such as that of the vagina.

The ejaculatory duct is the approximately one inch long tube of vas deferens and seminal vesicle duct joined together. The ejaculatory ducts lie within the prostate gland and empty their secretions into the urethra. The urethra is a single tube that originates at the bottom of the bladder and functions to carry both urine and sperm to the outside of the body in males. There are sphincter muscles around the urethra, above the ejaculatory ducts. During sexual intercourse these sphincters close to prevent urine from mixing with sperm.

The prostate gland completely surrounds the urethra just under the bladder. The prostate gland is normally about the size of a walnut, and it secretes a thin, milky fluid into the urethra. This fluid contains citrate which is also used by the sperm to make ATP. Older men may have prostate problems when the prostate becomes enlarged and squeezes the urethra shut so the bladder cannot be emptied. Often surgical removal of the gland is necessary to correct the problem.

The Cowper's glands (or bulbourethral glands) are about the size of a pea and lie on each side of the urethra below the prostate. They secrete an alkaline mucous which protects the sperm from acid and from being engulfed by white blood cells in the vagina and uterus. This mucous is also thought to be a precoital lubricant. The Cowper's glands secrete directly into the urethra.

The penis consists of a body or shaft which terminates in an extended area called the glans penis. The shaft is composed of three cylindrical bodies of erectile tissue. The single corpus spongiosum directly surrounds the urethra. The paired corpus cavernosa are on the dorsal side and make up most of the penis.

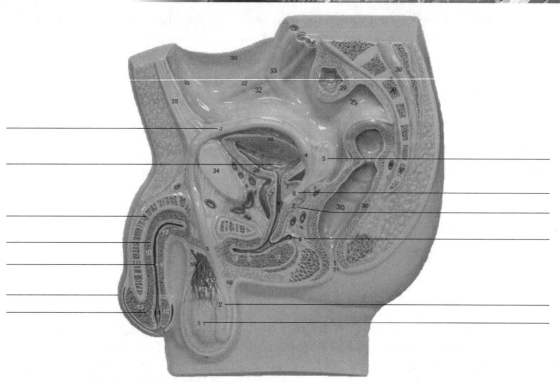

FIGURE 17.1   Male Reproductive Structures, Sagittal View
By Hubbard Scientific, www.amep.com. Reprinted by permission.

The glans penis is actually the enlarged end of the corpus cavernosa. Skin is loosely attached to the shaft and covers the glans as free flap. This flap, called the prepuce (or foreskin), is often removed surgically in the hospital shortly after birth. This surgery is called circumcision. Sexual stimulation causes the corpus spongiosum and corpus cavernosa to become engorged with arterial blood while the veins draining the penis are compressed so that very little blood leaves the area. This causes the penis to become and remain erect.

Semen is sperm plus all the secretions of the epididymis, seminal vesicles, prostate gland, and Cowper's glands. The epididymis contributes a small amount of fluid. The seminal vesicles contribute about 60% of the fluid to semen, the prostate contributes about 30% of the fluid, and the Cowper's glands release the remainder of the fluid. Each ejaculation contains about 2–5 mL of semen containing 20–150 million sperm. If the sperm count drops too low or contains too many malformed or nonfunctional sperm, the man is said to be sterile. This is because the higher the number of functional sperm, the higher the chances of one sperm surviving to fertilize an egg. Sperm also secrete hyaluronidase, an enzyme that breaks down the outer layer of follicular cells surrounding the egg. If the sperm count is too low, there will not be enough hyaluronidase released to allow even one sperm to penetrate the egg.

The female reproductive system starts in the paired ovaries, which are the sites of gamete production. The female gamete is the oocyte or egg. Like the testes, the ovaries are surrounded by tunica albuginea. The ovaries, and other female reproductive organs, are held in place by ligaments: suspensory ligaments, broad ligament, round ligament, and others.

Look at slide of a cat ovary in cross section. Embedded in the connective tissue of the ovary are numerous tiny sac-like structures called follicles. Each follicle contains an oocyte. There are numerous small primary follicles which have 1–2 layers of cells surrounding the oocyte. Primary follicles resemble small wagon wheels. Medium-sized developing follicles have a small fluid filled space (the antrium) between the oocyte and the surrounding cells. The biggest follicles are the mature Graafian follicles in which the oocyte sits on a stalk of cells positioning the oocyte close to the middle of the large antrium.

Each month, a Graafian follicle ruptures and releases the oocyte in a process called ovulation. A corpus luteum is a large, empty follicle after the egg has ruptured from the follicle. The corpus luteum has convoluted walls that are much thicker than the walls of the Graafian follicles and would lack the oocyte.

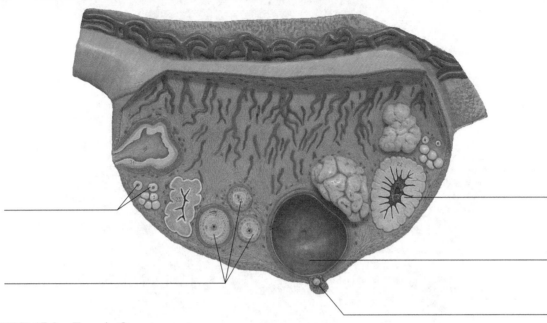

FIGURE 17.2   Female Ovary
SOMSO Model of Ovary.

In the space below, draw and label a graafian follicle and corpus luteum as seen under the microscope.

The **fimbriae** are the finger-like projections of the fallopian tubes. The function of the fimbriae is to sweep the egg into the fallopian tube. The **fallopian tube** (or oviduct or uterine tube) leads from the fimbriae to the upper portion of the uterus.

The **uterus** or womb is the organ which nurtures the developing child during pregnancy. The pear-shaped uterus lies behind the bladder and in front of the rectum inside the pelvic cavity. The innermost layer of the uterus is the **endometrium**. Following the birth of a baby or during menstruation, part of the endometrium is sloughed off. The thick middle layer of the uterine wall is the **myometrium**, which is composed of smooth muscle fibers. This gives great strength to the uterine wall and allows it to contract during childbirth. The thin outermost layer of the uterus is the **perimetrium**. The **cervix** is the neck of the uterus and is located at the lower end of the uterus. During a **Pap smear** some of the cells of the cervix are scraped away; these cells are later tested for cervical cancer.

The **vagina** is a muscular canal, four to six inches in length, leading from the cervix to the exterior. Being composed mainly of smooth muscle lined with mucus membranes, the vagina is capable of great distention, such as during childbirth. The vagina contains many mucous glands which secrete a lubricant during sexual stimulation. The pH of the vagina is normally acidic, which kills many microorganisms.

The external female genitalia are collectively referred to as the **vulva**. The **mons pubis** is a skin-covered mound of fatty tissue over the area where the pubic bones join in front. Below the mons pubis is the **clitoris**, a small organ of erectile tissue. The clitorus, like the penis (its homologue), becomes erect due to engorgement with blood during sexual stimulation. The clitoris also has an enlarged glans that is extensively innervated. The **labia majora** are the larger two longitudinal folds of adipose tissue covered

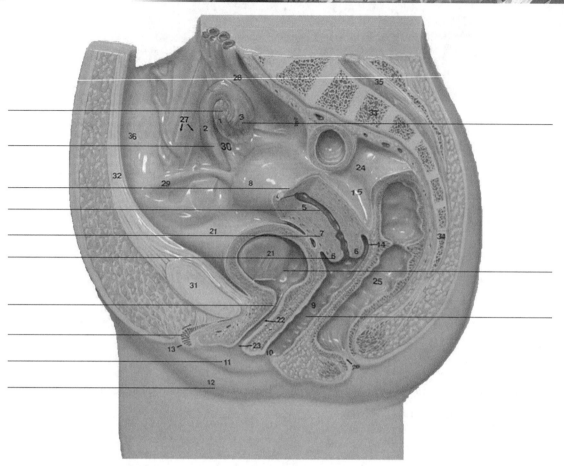

FIGURE 17.3   Female Reproductive Structures, Sagittal View
By Hubbard Scientific, www.amep.com. Reprinted by permission.

with pigmented skin. The labia majora are the homologues of the scrotum. The smaller inner folds of skin are called the labia minora and surround the vestibule or entrance to the vagina.

The female urethra originates at the bladder and carries only urine to the exterior. It is not considered to be a reproductive organ.

Oogenesis is the production of eggs. Oogenesis starts very early in females—before birth! In the ovaries, primary oocytes are produced by mitotic division. The primary oocyte matures in the follicle, then undergoes the first meiotic division and forms a polar body. A polar body is a nonfunctional egg and will rapidly disintegrate. The primary oocyte will stall in a state of suspended animation until needed. After puberty each ovary normally selects one oocyte to reactivate approximately every two months.

The randomly selected oocyte will become a primary follicle, developing follicle, Graafian follicle, and undergo ovulation. The oocyte will be swept into the fallopian tube and moved toward the uterus by ciliary action.

If the egg meets a sperm, the oocyte will undergo the second meiotic division giving the egg the correct amount of DNA. A second polar is also produced and it degenerates. The sperm penetrates the secondary oocyte, the two nuclei fuse, and the fertilized egg is now called a zygote, or one-celled embryo. Fertilization, the union of the sperm and egg, occurs in the first third of the fallopian tube. The zygote will undergo mitotic divisions, become a hollow ball of cells, and implant into the endometrium of the uterus. Through the hormonal action of progesterone, normally there will be no menstruation during pregnancy.

However, if the oocyte does not meet a sperm in the fallopian tube within a few hours of ovulation, the oocyte disintegrates, is absorbed by the body, and rarely reaches the uterus. About two weeks after ovulation, the endometrial lining is sloughed off. The debris is passed out of the uterus and vagina. This process is called menstruation. The endometrium at once begins building a new lining in preparation for the next ovulation. About two weeks after menstruation, ovulation occurs again, and if no fertilization takes place, the cycle on menstruation and ovulation is repeated.

Name: _____

QUESTIONS

1. Name the male and female gametes and the process to make each.

2. Semen consists of sperm plus secretions from four reproductive structures. Name the structures.

3. What is the function of interstitial cells? Where are they located?

4. Which is larger, a sperm or an egg? Which is motile?

5. Can the removal of the prostate gland cause sterility? Why or why not?

6. What part does the male urethra play in reproduction?

7. What part does the female urethra play in reproduction?

8. What is the function of the endometrium?

9. A Graafian follicle and a corpus luteum are usually about the same size under the microscope. What is the most obvious difference between the two?

10. If each ovary normally selects one oocyte to reactivate approximately every two months, why do women have periods every month?

18 LABORATORY

Human Development

A. Instructional Objectives

After completing this lab exercise the student should be able to
1. Identify where fertilization takes place.
2. Trace the development of the zygote sequentially through embryonic stages.
3. Describe the fate of embryonic structures and what they will become.
4. Identify all structures and know their function.

B. Introduction

Fertilization refers to the process of a sperm fusing with an egg (technically a secondary oocyte). Although "pregnancy" should start with fertilization, the gestation period officially starts from the date of the woman's last period. Thus a woman is technically 2 weeks pregnant at fertilization! The term embryo refers to the first two months after fertilization. Starting at the beginning of the third month, the developing baby is referred to as a fetus.

After the union of the sperm and egg, the resulting embryo/fetus will spend approximately 266 days in the mother's uterus. During this time the single microscopic cell (zygote) will be transformed into a trillion-celled being capable of carrying on all life processes on its own. The almost miraculous changes that take place during this developmental phase of life are topics of this lab exercise.

C. Materials
Drawings, text, developmental models, and wall charts
Slide of sea urchin development

D. Methods

Each ovary will release a secondary oocyte every other month. The oocyte is caught by the fimbriae and transported to the fallopian tube. Meanwhile, during ejaculation millions of sperm are deposited in the vagina. Although the egg is only about 5 inches away in a fallopian tube, relatively few of the sperm will reach the egg. Most of the sperm leak out of the vagina; millions of sperm are destroyed by the acidity of the vagina; most sperm do not find their way through the cervix to enter the uterus. Half of the remaining sperm swim towards the fallopian tube without the egg. Only a 100–1000 sperm will reach the oocyte in the upper third of the fallopian tube.

The oocyte is surrounded by a layer of cells called the corona radiata. The acrosome of the sperm contains an enzyme called hyaluronidase, which chemically digests through the corona radiata. Although it is said that "it only takes one sperm to fertilize an egg," it generally takes hundreds of sperm releasing their hyaluronidase to fully digest through the egg's corona radiata.

Underneath the corona radiata, and directly surrounding the oocyte, is the zona pellucida or fertilization membrane. Once the first sperm reaches the zona pellucida, enzymes cause an almost instantaneous change in the zona pellucida which hardens and prevents more sperm from entering the egg. This prevents polyspermy or multiple sperm fertilizing one egg. If two sperm were to fertilize an egg, there would be too many copies of each chromosome and the resulting embryo would not be viable. The zona pellucida remains until the growing mass of cells reaches the uterus.

Once the sperm enters the secondary oocyte, the sperm loses its midpiece and tail. The secondary oocyte undergoes the second round of meiosis to form an ovum or egg. The sperm's DNA mixes with the egg's DNA to form a zygote, a one-celled embryo, with the correct number of chromosomes.

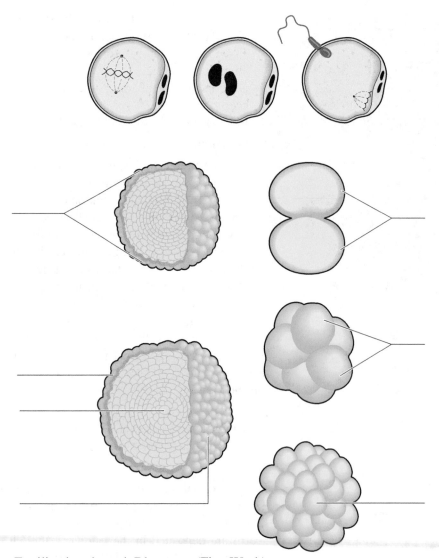

FIGURE 18.1   Fertilization through Blastocyst (First Week)
Drawing provided by Jerry D. Barton, II M. S. with permission.

Within hours of fertilization, the zygote begins cleavage, or mitotic divisions without cell growth. The one-celled zygote will divide into a 2 cell embryo. While the 2 cells are each about half the size of the original cell, due to cleavage the overall size of the embryo is the same. The 2 cells will divide into the 4 cell stage and then the 8 cell stage. Approximately 3 days after fertilization, a ball-like cluster of cells called a morula (consisting of 16–64 cells) has been formed. A blastocyst (or blastula) is a hollow ball of hundreds of cells. Although each of the cells is tiny, the overall size of the embryo is still approximately the same size as the original zygote. Each individual cell of a morula or blastocyst is called a blastomere.

How many blastomeres are there in a morula? _____

The blastocyst has a hollow center with a cluster of cells off to one side. The hollow center is referred to as the blastocoel. The cell cluster is called the inner cell mass. The inner cell mass will form the embryo itself, plus several extraembryonic membranes found outside the embryo. The single layer of flat cells around the outside of the blastocyst is the trophoblast. The trophoblast forms the chorion and later the placenta.

Examine a sea star development slide under the microscope. Identify a zygote, 2 cell stage, 4 cell stage, 8 cell stage, morula, and blastocyst. Draw each stage in the space below.

It takes three to five days from the time of ovulation for the mass of rapidly dividing cells to travel the length of the fallopian tube and reach the uterus. In the uterus, the embryo finally becomes too big for the zona pellucida to hold all the cells, and the embryo "hatches" out of the zona pellucida. The mass of cells may remain in the uterus for several days before it rests on the endometrium of the uterus and embeds itself in the spongy, vascular tissue—implantation. By the end of the first week, the blastocyst becomes firmly implanted within the endometrium. The trophoblast begins to become thicker and form numerous root-like structures which start to grow into the surrounding endometrium. These structures called villi (villus - singular, villi - plural) contain tiny cavities which fill with the mother's blood. The embryo's blood vessels will develop later and absorb nutrients from the mother's blood by means of diffusion and active transport. Waste materials are passed from embryo to mother by the same way.

About midway through the second week, changes begin to appear within the inner cell mass. At first the cells are all very similar, but soon they begin to differentiate and form distinctly different cell layers. The cells nearer to the trophoblast appear larger and form the first of three primitive germ layers. The ectoderm, or outer germ layer, will form structures of the nervous system and the epidermis of the skin. The endoderm, or inner germ layer, will form the epithelial linings of the digestive, respiratory, and urogenital systems, associated glands, liver, and yolk sac. The mesoderm is the third and last of the primary germ layers to appear; it forms muscles, bones, blood, and other remaining body parts. This process of differentiation of the inner cell mass is called gastrulation. During approximately the second week, the developing embryo is referred to as a gastrula. Label Figure 18.2 Early Gastrula.

Meanwhile the four extraembryonic membranes start to form. The amnion will completely enclose the embryo, broken only by the umbilical cord. The amnion will become filled with fluid and is later called the "bag of waters" that breaks late in pregnancy. The fluid-filled amnion protects the developing embryo from physical trauma and helps maintain homeostasis. When *amniocentesis* is performed, a needle is inserted through the mother's abdomen and uterus, into the amnion, and some of the amniotic fluid is removed for genetic testing.

While in many species the yolk sac provides nutrients to the developing embryo, in humans the yolk sac produces the earliest blood cells and blood vessels. The allantois is found in the umbilical cord and aids in the transfer of material between mother and fetus. The chorion is the outermost membrane and is

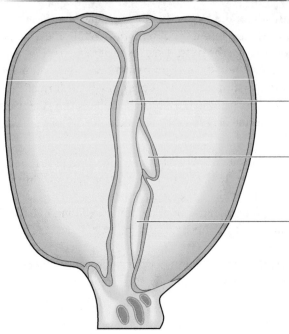

FIGURE 18.2 Early Gastrula (Second Week)
Drawing provided by Jerry D. Barton, II M. S. with permission.

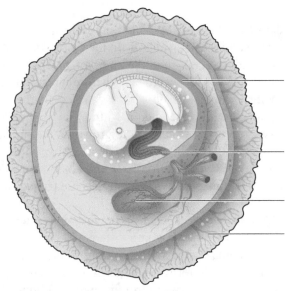

FIGURE 18.3 Extraembryonic Membranes (At Nine Weeks)
Drawing provided by Jerry D. Barton, II M. S. with permission.

formed by the trophoblast. The chorion will help form the placenta; it encloses the embryo and all other membranes. When fully formed the amnion, yolk sac, and embryo are attached to the chorion by a short segment of the epithelial linings of the digestive, respiratory, and urogenital systems called the body stalk. The **body stalk** later develops into the umbilical cord. Label Figure 18.3 Extraembryonic Membranes.

During the third week, the embryo itself takes shape, though not a human shape. A **primitive streak** appears as a line or groove along the dorsal surface of the embryo and establishes the longitudinal axis of the body. Mesodermal cells form rectangular blocks called somites. **Somites** form bone and muscle; the somites along the primitive streak will form the bones and muscles of the spine. Endodermal cells on the inside of the embryo form the **primitive gut** or first digestive tract. Ectodermal cells along the

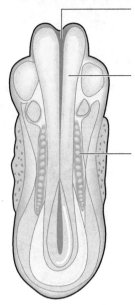

FIGURE 18.4 Neurula (Third Week)
Drawing provided by Jerry D. Barton, II M. S. with permission.

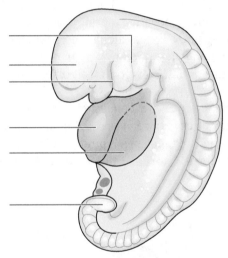

FIGURE 18.5 Four Week Old Embryo
Drawing provided by Jerry D. Barton, II M. S. with permission.

primitive streak form the neural tube, which later becomes the brain and spinal cord. This process of developing nervous tissue is called neurulation. The embryo during the third week is called a neurula. Label Figure 18.4 Neurula.

Organogenesis is the process of making organs. The brain and spinal cord are the first organs to begin development. The heart is also formed late in the third week and is the first organ to function, since it begins beating almost as soon as it is formed.

During the fourth week other internal organs begin forming—the liver, kidneys, lungs, major blood vessels, etc. Muscles begin appearing in the trunk and cartilage begins to appear where the backbone will later be found. A series of bulges and slits appear in the facial and neck region. They will form the mouth and jaws (mandibular arch), ear (hyomandibular cleft), and part of the larynx (pharyngeal gill cleft). The optic vesicle becomes apparent where the eye will form, but facial features are distinctly nonhuman at this time.

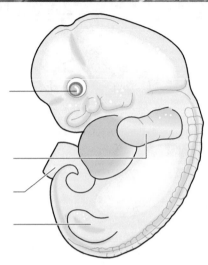

FIGURE 18.6   Five Week Old Embryo
Drawing provided by Jerry D. Barton, II M. S. with permission.

The **heart** and **liver** are conspicuous bulges or protuberances. The number of somites has increased to thirty and the embryo has a distinct **tail** at this stage. The amnion has grown immensely and now completely surrounds the embryo. Label Figure 18.5 Four Week Old Embryo.

The most conspicuous features appearing during the fifth week are the first signs of limbs. Swellings, the **arm buds** and **leg buds**, appear along the vertebral column which is still composed entirely of cartilage. The head and brain are further along in development than most other organs. Facial features are continuing to change as nasal pits form well out towards either side of the face. **Nasal pits** will become the nostrils.

The **umbilical cord** and **placenta** are connected via blood vessels and begin to function in exchanging food, oxygen, and wastes between the developing embryo and mother. The yolk sac is virtually useless now as the growing liver and spleen produce blood cells for the embryo. The allantois in the umbilical cord is fully functional. Label Figure 18.6 Five Week Old Embryo.

FIGURE 18.7   Nine Week Fetus
Drawing provided by Jerry D. Barton, II M. S. with permission.

During the sixth through eighth weeks, the head and facial features become greatly refined. The nasal pits move from the sides of the face to meet in the middle and form the nose. Two bulges, called the maxillary processes, begin on the sides of the head, grow to meeting the midline of the face, and form the upper jaws. The mandibular arch forms the lower jaw, thus producing a mouth. The hyomandibular clefts form the auditory canals. Bulges above this cleft begin to form and ultimately make the external ear flaps as the ear begins to shift upward to a position of eye level. The eyes are formed but are not refined and have no lids.

The reproductive organs begin rapid development early in the second month. By the end of the eighth week the sex of the child can be determined from the embryo, though the organs are not completely developed yet. Limb buds develop paddle-like hands and feet. These paddles will become differentiated into toes and fingers during the last part of the second month.

By the end of eighth week, essentially everything is there. All the organs are formed. Cartilage is found in all places of the skeleton where bones will be found later. The embryo is still only about one inch long and weighs less than one ounce and the mother has not yet begun to show her pregnancy.

The beginning of the third month marks the end of the embryo and the beginning of the fetus. The terminology changes to "fetus" after all of the organs are formed and bone begins to be deposited within the cartilage skeleton. Teeth begin their development and all internal organs, with the exception of the lungs and reproductive organs, are functioning. The fetus does not yet crowd the amnion and the chorion is still intact. Later it tears away as the fetus gets too large for it. The amniotic fluid acts as an insulator and shock absorbent for the fetus. The heart beat can sometimes be picked up by a stethoscope during the third month.

Months three through nine are primarily months of further growth and refinement. Eyelids, eyebrows, lips, nails, fingerprints, and footprints appear. A waxy protective covering is secreted by the oil glands. The fetus rapidly increases in length and weight. By the end of the fourth month, it is 10 inches long and weighs about half a pound. By the sixth month, it is 14 inches long and about 2 pounds.

By the end of the seventh month, the fetus has a reasonable chance of survival if born, providing an incubator and medical attention is available. About 50% of the birth weight is gained during the last two months. At birth, the average infant is between 18 and 21 inches in length and weighs between 6 to 8 pounds.

Summary of Development

Week One

Ovulation

Fertilization in fallopian tube

Cleavage of zygote

Formation of morula

Formation of blastocyst

Implantation

Week Two

Appearance of chorionic villi

Differentiation of inner cell mass

Formation of ectoderm, endoderm, and mesoderm

Amnion begins to form

Week Three

Formation of body stalk

Formation of primitive streak

Formation of neural tube

Appearance of first somites

Appearance of heart

Amnion nearly covering embryo

Chorion well developed

Week Four

Heart begins functioning

Brain and spinal cord begin forming

Internal organs begin forming

All somites present

Amnion fully covers embryo

Facial features and ears appear

Muscles and cartilage skeleton appear

Second Month

Internal organs completed with some functioning

Muscles completed and cartilage skeleton completed

Arms, legs, and digits develop from limb buds

Ears and facial features are nearly complete

Embryo becomes distinctly human and appearance

Placenta and umbilical cord becomes functional

Third through Ninth Month

Embryo now technically a fetus

Bone formation within cartilage skeleton begins

External genitalia becomes differentiated

Nails, hair, eyebrows, and skin features develop

Growth and refinement of detail

Organs continue rapid development

Skeletal system becomes mostly replaced by bony tissue

Respiratory and reproductive structures are not fully functional until after birth.

Name: _____

QUESTIONS

1. Briefly describe what happens to the fertilized egg as it travels through the fallopian tube.

2. Arrange the following stages or events in their correct sequence.
   a. blastocyst
   b. fetus
   c. gastrulation
   d. inner cell mass
   e. limb buds
   f. morula
   g. neural tube formation
   h. organ formation
   i. somites appear
   j. umbilical cord and placenta functioning

   1. _____
   2. _____
   3. _____
   4. _____
   5. _____

   6. _____
   7. _____
   8. _____
   9. _____
   10. _____

3. Match up the two columns. On the left is a structure which later gives rise to or forms the structure on the right.

   1. body stalk _____
   2. cartilaginous skeleton _____
   3. hyomandibular cleft _____
   4. inner cell mass _____
   5. mandibular arch _____
   6. maxillary process _____
   7. neural tube _____
   8. nasal pit _____
   9. somites _____
   10. trophoblast _____

   a. bony skeleton
   b. ear
   c. embryo
   d. lower jaw
   e. nostril
   f. placenta
   g. spinal cord, brain
   h. umbilical cord
   i. upper jaw
   j. bone and muscle

4. List some of the major structures which arise from each of the primitive germ layers.

   Ectoderm—

   Mesoderm—

   Endoderm—

5. Explain how the exchange of oxygen, carbon dioxide, food, and wastes take place between the mother and fetus via the placenta, even though the blood of each does not actually intermingle.

6. At the end of which month does the fetus have all the essential organs? What size is the fetus at this point?

7. What is the function of the yolk sac in humans?

8. What organ is the first to function? Which two organ systems do not function until after birth?

9. At what month can the sex of the developing fetus be determined visually?

10. At what point in the pregnancy does the fetus have a chance of survival if born prematurely?

11. Although the fetus develops in the uterus, where does fertilization take place?

19

DNA, RNA, and Protein Synthesis

A. Instructional Objectives: After completing this exercise the student should be able to:
1. Identify the following in a model of DNA:
   a. Nitrogenous bases
   b. Deoxyribose sugar
   c. Phosphate group
2. Identify the following in a model of RNA:
   a. Nitrogenous bases
   b. Deoxyribose sugar
   c. Phosphate group
3. Explain the meaning of the following terms:
   a. Nucleotide
   b. Codon
   c. Anticodon
   d. Triplet
   e. Double helix
   f. Complementary strand
4. Know the abbreviations for the following nitrogenous bases:
   a. A—Adenine
   b. C—cytosine
   c. G—guanine
   d. T—thymine
   e. U—uracil
5. Distinguish between a DNA molecule and an RNA molecule.
6. Know the site of formation and explain the functions of the following types of RNA:
   a. Messenger RNA (mRNA)
   b. Transfer RNA (tRNA)
   c. Ribosomal RNA (rRNA)
7. Form the following:
   a. A complementary strand of DNA from a parent strand of DNA
   b. A complementary strand of mRNA from a parent strand of DNA
   c. A chain of amino acids from an mRNA

8. Distinguish between the following:
    a. A purine and pyrimidine nitrogenous base
    b. Normal hemoglobin, hemoglobin A (Hb^A) and sickle cell hemoglobin, hemoglobin S (Hb^S)
    c. Replication, translation, and translation
    d. Point mutation and a frame shift mutation
9. Explain how a mutation occurs.
10. Identify a normal RBC and an RBC with sickle cell anemia.
11. Isolate DNA from the strawberry.

B. Materials

DNA model

Microscopes

Slides with normal red blood cells (RBCs)

Slide with sickle cell anemia

1–3 strawberries (about the volume of a golf ball; frozen strawberries should be thawed at room temperature)

15-mL DNA extraction buffer

About 20 mL ice cold 97% ethanol

1 heavy-duty Ziploc freezer bag

1 small clear test tube

1 funnel

Filter paper

1 transfer pipe

C. Reagents

DNA extraction buffer, 1 L: Mix 100 mL of shampoo (without conditioner), containing Laurel sulfate (SDS, 15 g NaCl [sea salt], 60 g papain, 900 mL water) or 50 mL of detergent containing Laurel sulfate (SDS, 15 g NaCl, 900 mL water).

Papain is the enzyme that will break down the structural proteins and other enzymes. Papain is found in certain brands of meat tenderizers.

Soaps that work well: Lemon Fresh Joy, Woolite, Ivory, Shaper, Arm & Hammer, Herbal Essence shower gel by Clairol, Tide, Dish Drops, Kool Wash, Cheer, Sunlight Dish Soap, Dawn, Delicate, All, and Ultra Dawn.

D. Introduction

The two types of nucleic acids are DNA and RNA. DNA is the molecule that makes up our genes located on chromosomes in the nucleus of our cells. Your height, intelligence, and eye, skin, and hair color are determined by your genes. Your DNA gives your cells the instructions for making the various proteins necessary for proper functioning of the cell. Table 19.1 lists examples of different classes of common proteins found in animals.

TABLE 19.1   Classes of Proteins in Animals

Class of Protein	Example	Function
Enzyme	Lactase	Act as catalyst; controls the rate of chemical reactions
Hormone	Insulin	Chemical messengers in the blood that regulate metabolism
Contractile	Actin	Movement of muscles and cell parts
Carrier	Hemoglobin	Transports materials throughout the body
Globulins	Antibodies	Protects the body from infection
Toxins	Venoms	Animals use these proteins to protect themselves or capture prey

TABLE 19.2   Classes of RNA

Class	Abbreviation	Function
Messenger RNA	mRNA	Carries the code for a protein-coding gene from DNA to ribosomes
Ribosomal RNA	rRNA	Combines with proteins to form ribosomes, the stuctures that link amino acids to form a protein
Transfer RNA	tRNA	Carries the proper amino acid to the ribosome

The other nucleic acid is RNA, which is a formed in the nucleus of the cell. There are three classes of RNA: messenger RNA (mRNA), ribosomal RNA (rRNA), and transfer RNA (tRNA). All three types of RNA are directly involved in the synthesis of proteins in the cytoplasm of the cell. Messenger RNA is formed in the nucleus and travels to the ribosome. Ribosomal RNA is found in the ribosomes. Transfer RNA is a cloverleaf-shaped molecule found in the cytoplasm that is involved in bringing the appropriate amino acid to the ribosome.

The shape of a DNA molecule is described as a double helix. This molecule is also described as a spiral staircase. The rungs of the spiral staircase are the nitrogen bases while the backbone is made of sugar (deoxyribose) and phosphates. Four types of nitrogen bases are found in DNA: adenine (A), thymine (T), cytosine (C), and guanine (G). Adenine and guanine are double-ring nitrogen bases called **purines**. Cytosine and thymine are single-ring nitrogen bases called **pyrimidines**. Adenine always base pairs with thymine, and cytosine always base pairs with guanine. The base pairs are held together by hydrogen bases. Hydrogen bonds are weak intermolecular bonds between a hydrogen atom and an oxygen or nitrogen atom. The bond strength of a hydrogen bond is 1/20th the strength of a covalent bond.

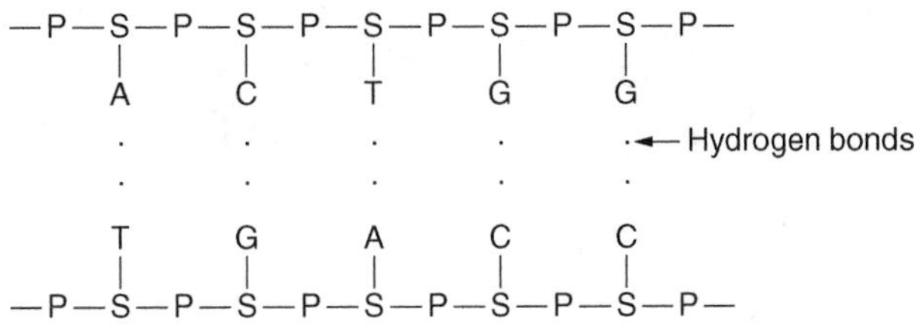

FIGURE 19.1   Base-Pairing of DNA

RNA is a single-strand molecule. RNA is composed of four different nitrogen bases. Three of these are the same as DNA (adenine, guanine, and cytosine). Instead of thymine, we find the pyrimidine uracil. In RNA as in DNA, cytosine always base pairs with guanine. However, there is a different base pairing with adenine. In RNA adenine always base pairs with uracil.

TABLE 19.3   Comparison of DNA and RNA

Nucleic Acid	DNA	RNA
Shape of Molecule	Double helix	Single Strand
Type of Sugar	Deoxyribose	Ribose
Unique Nitrogen Base	Thymine	Uracil
Function	Contains genes; sequence of bases determines the amino acid sequence of a protein	Involved in gene expression; transcription and translation

When a protein is to be synthesized in the cell at the ribosome, the segment of the DNA that encodes for that protein unzips. By the aid of the enzymes, mRNA is synthesized. This process is called **transcription**.

```
        —P—S—P—S—P—S—P—S—P—  ◄— DNA strand
            |     |     |     |
            A     C     T     G
            .     .     .     .  ◄— Hydrogen bonds
            .     .     .     .
            U     G     A     C
            |     |     |     |
        —P—S—P—S—P—S—P—S—P—  ◄— mRNA strand
```

FIGURE 19.2   Transcription of DNA

Proteins are made of a chain of 20 different naturally occurring amino acids. If the genetic code was one nitrogen base for each amino acid, there could be only four amino acids in proteins. The problem is that in nucleic acids, there are only four different nitrogen bases. Proteins are made up of "words" called amino acids. The language of nucleic acids is also made of "words" called **codons**. Each codon is a triplet of nitrogen bases. There are 64 different codons possible and there are 20 different amino acids. This means that some of the codons are used over to mean the same amino acid. This process of changing from the language of nucleic acids to the language of proteins is called **translation**. The site of translation is the ribosome. Ribosomes are granular organelles that are found in all cells where protein synthesis occurs.

```
                        AA ◄— amino acid
                        |
            S—P—S—P—S—
            |     |     |
            A     C     U ◄— anticodon (tRNA molecule)
            .     .     .
            .     .     .
            U     G     A
            |     |     |
          —P—S—P—S—P—S— ◄— codon (mRNA strand)
```

FIGURE 19.3   Translation of mRNA

It is the sequence of the nitrogen bases that determines the sequence of the amino acids in the protein. Sometimes mistakes are made in the DNA, called **mutations**. If there is the **inversion** or **substitution** of a nitrogen base, the type of mutation that occurs in this is called a **point mutation**. If a point mutation occurs in the key part of the code for a protein molecule, it can make the molecule dysfunctional. Let us use the sentence "The cat ate the rat" as an example of a normal protein. Each word represents an amino acid. An example of an inversion point mutation would be the sentence "The cat are the rat." A substitution point mutation would be "The dog ate the rat." As you can see, the whole meaning of the sentence, the protein, can be changed by one word, an amino acid.

There are two **frame-shift mutations**, deletions and insertions. If a nitrogen base is lost, then this is called a **deletion**. Using our sentence "The cat ate the rat," in a deletion it would result in "The cat the rat," which does not make much sense. If a nitrogen base is added, then this is called an **insertion**. Again with our simple sentence, if an amino acid (word) is added such as an extra "the": "The the cat ate the rat." When frame-shift mutations occur at the beginning of the strand of DNA coding for a protein, the entire protein can be nonfunctional or dysfunctional.

Most mutations are lethal—that is, the individual dies before sexual maturity. However, some rare mutations are beneficial. These may be passed on to future generations. Mutations are spontaneous. They may occur at any time. Sickle cell hemoglobin is due to a point mutation.

The language of proteins is determined by the sequence of 20 naturally occurring amino acids. This sequence of amino acids is called the **primary structure** of a protein. It is the primary structure that ultimately determines the shape and the function of the protein. In biology, shape determines function. Sometimes if an amino acid is omitted or is substituted with another amino acid, and this occurs in a key place in the chain, the molecule can become nonfunctional or dysfunctional. For instance, there are two types of hemoglobin possible in the human body, hemoglobin A, Hb^A, and sickle cell hemoglobin, Hb^S. Later in this laboratory we will examine the genetic difference between these different hemoglobin molecules. Every protein that is formed in your body has its instructions encoded in the DNA in the nucleus of each cell.

In this laboratory we will simulate protein synthesis by forming sentences from a strand of DNA. Protein synthesis consists of two phases: transcription and translation. Transcription is like the changing of dialects; one dialect is DNA and the other dialect is RNA. The dialect of DNA is changed into the dialect of RNA. The reasons for this change in dialect are that DNA is a macromolecule too large to pass through the pores of the nucleus, and if DNA did move out of the nucleus, the DNA might be lost for future generations of cells. To understand DNA, we first need to review the structure of DNA.

Part 1: Mutations

Procedure A. The DNA Model

1. Using the DNA model identify the following nitrogenous bases:
   A: _____
   C: _____
   G: _____
   T: _____

2. To which nitrogenous base is the following always paired?
   Adenine: _____
   Thymine: _____

3. Identify the bond represented by type of chemical bond between the two nitrogen bases.
4. What is the name of the sugar found in DNA?
5. Which nitrogenous bases are purines?
6. Which nitrogenous bases are pyrimidines?
7. List two reasons you know the model of a section of DNA and not RNA.
   a. _____
   b. _____

8. The following is the nitrogenous base sequence on one strand of a DNA molecule. Write the base sequence of the complementary strand of DNA.

Strand	A C G	C C A	G T G	G G T	T C G	A T C
Complementary Base Strand						

Procedure B. Transcription and Translation

1. From the DNA strand base sequence, write the complementary mRNA strand.
2. Using the handout of mRNA codons, determine the sequence of amino acids in the polypeptide.

DNA Strand Base Sequence	ATG	G C C	A G T	G G T	T C G	C A C C C
mRNA Strand Base Sequence						
Amino Acid Sequence						

Procedure C. Change in the Sequence of Nitrogenous Bases

1. The following is the base sequence on one strand of DNA molecule:

Original DNA Strand Base Sequence	A T G	C G C	A G T	G G T	T C G	C A C C C

2. The new DNA strand base sequence is formed when the fourth nucleotide is changed from G to C.
3. Write the base sequence of the strand mRNA transcribed parent DNA strand.
4. From the handout, write the sequence of amino acids in the polypeptide.

Original DNA Strand Base Sequence	A T G	C G C	A G T	G G T	T C G	C A C C C
New DNA Strand Base Sequence						
mRNA Strand Base Sequence						
Amino Acid Sequence						

QUESTIONS

1. Did a mutation occur? _____
2. If so, what type? _____
3. Was there a change in the sequence of amino acids in the polypeptide? _____

Procedure D. Addition of a Nitrogen Base in a DNA Sequence

1. The following is the base sequence on one strand of DNA molecule.

Original DNA Strand Base Sequence	A T G	C G C	A G T	G G T	T C G	C A C C C

2. In this nitrogenous base sequence, change the fourth nucleotide in the parent DNA strand by adding G after the third nucleotide.
3. Write the base sequence of the strand mRNA transcribed parent DNA strand.
4. From the handout, write the sequence of amino acids in the polypeptide.

Original DNA Strand Base Sequence	A T G	G C C	A G T	G G T	T C G	C A C
New DNA Strand Base Sequence						
mRNA Strand Base Sequence						
Amino Acid Sequence						

QUESTIONS

1. Did a mutation occur? _____
2. If so, what type? _____
3. Was there a change in the sequence of amino acids in the polypeptide? _____

Procedure E. Addition of a Nitrogen Base in a DNA Sequence

1. The following is the base sequence on one strand of DNA molecule.

Original DNA Strand Base Sequence	A T G	C G C	A G T	G G T	T C G	C A C C C

2. In this nitrogenous base sequence, change the eighth nucleotide in the parent DNA strand from G to C. Write the *new* DNA strand sequence.
3. Write the base sequence of the strand mRNA transcribed parent DNA strand.
4. Using the tRNA codons and amino acids handout, write the amino acid sequence for the polypeptide.

Original DNA Strand Base Sequence	A T G	G C C	A G T	G G T	T C G	C A C
New DNA Strand Base Sequence						
mRNA Strand Base Sequence						
Amino Acid Sequence						

QUESTIONS

    1. Did a mutation occur? _____

    2. If so, what type? _____

    3. Was there a change in the sequence of amino acids in the polypeptide? _____

Part 2. Varieties of Hemoglobin

Hemoglobin (Hb) is a transport molecule for oxygen that is found in red blood cells in humans. The hemoglobin molecule consists of two alpha chains and two beta chains. Each hemoglobin molecule contains four iron-containing heme groups. There are thousands of hemoglobin molecules in each red blood cell. The oxygen-carrying capacity of hemoglobin is determined by the shape of this molecule. A distorted shape of hemoglobin will lower its oxygen-carrying capacity. There are two varieties of hemoglobin in the equatorial regions of Africa: hemoglobin A (HbA) and hemoglobin S (HbS). The equatorial areas of Africa are ravaged by the disease malaria. Malaria is a very serious disease that kills 1 million individuals worldwide each year. These individuals are mostly poor children. About 4000 years ago, before the invention of insecticide DDT and the drug chloroquine to combat the malaria, the sickle cell hemoglobin gene appeared in this region of Africa due to a point mutation. This gene is still present in the gene pool today because of malaria. The parasite that is responsible for malaria is called *Plasmodium*. It lives part of its life cycle in the red blood cells. The other part of its life cycle is lived within the female *Anopheles mosquito*. When an individual is bitten by a female *Anopheles* mosquito infected with this parasite, it must complete its life cycle in red blood cells. Here enters the reason why hemoglobin S is still present. Those individuals who have one allele for hemoglobin S and one allele for hemoglobin A are a little more resistant to the effects of malaria. Remember, it is the children who are most affected by the disease. This single allele gives the individual an opportunity to reach sexual maturity. The individuals who have one allele for hemoglobin S are said to have sickle cell trait. The red blood cells of those who have sickle cell trait do not live as long as the red blood cells of those who have hemoglobin A. Hemoglobin S distorts its shape when in under acidic conditions—that is, low-oxygen conditions in capillaries of the tissues. As a result, the cells become worn faster when traveling through the capillaries and are destroyed. The premature death of the red blood cells interferes with the life cycle of *Plasmodium. Plasmodium* has a 48- to 72-hour life cycle in the red blood cells. The parasites increase in number within a red blood cell and cause it to burst. The metabolic waste products are released when the red blood cell bursts. These waste products cause the chills and fever characteristic of malaria. If the life cycle is interrupted, then the effects of the disease are diminished. As a result, the individual lives longer. If an individual receives at fertilization two alleles for sickle cell hemoglobin, then that individual is said to have sickle cell anemia. This is a very serious and sometimes fatal genetic disease. The red blood cells of these individuals under acidic, that is, low-oxygen, conditions become deformed into what is called sickle shapes. The sickle-shaped red blood cells clog the capillaries to the brain and the extremities. This prevents oxygen from getting to the tissues. With modern medicine, there are drugs available to be administered to help alleviate the symptoms so that infected individuals can live longer and more productive lives.

Procedure:

1. Observe under high power a slide of normal human blood and a slide of blood from an individual who has sickle cell anemia. Make a sketch of each cell.

Sickle cell anemia   Normal red blood cells
red blood cells

2. Complete the table below with the correct codons. Refer to the handout of mRNA codons.

Amino Acid Sequence for Hemoglobin S	mRNA codon	mRNA codon	Amino Acid Sequence for Hemoglobin A
Meth			Meth
Val			Val
His			His
Leu			Leu
Thr			Thr
Pro			Pro
Glu			Val
Glu			**Glu**

a. What is the difference between the amino acid sequence of hemoglobin A and hemoglobin S?

b. What is the difference between the mRNA strands for each type of hemoglobin?

Part 3. Extraction of DNA from Strawberries

Most cultivated strawberries are polyploids. The prefix *poly-* means many, and *ploid* refers to sets of chromosomes. Polyploidy is very common in domestic plants, especially in angiosperms (the flowering plants). Polyploids have enlarged ovaries and showy flowers. Many of today's angiosperms are thought to be polyploid. The strawberry *Fragaria ananassa* is an octoploid with eight sets of chromosomes. With eight sets of chromosomes, the strawberry is an excellent source for DNA to be extracted.

Procedure:

1. If fresh strawberries, remove the green sepals from the strawberries.

2. Place three to four fresh strawberries or if frozen place a sample of thawed strawberries about the size of a golf ball into a heavy-duty Ziploc freezer bag.

3. Remove the air from the freezer bag and seal it shut.

4. Mash the strawberries for about five minutes. This is a mechanical breakdown of the cell wall and the cell structures. The longer the strawberries are mashed, the better your results will be.

5. Add 15 mL DNA extraction buffer (soapy salty water) and mash for a few more minutes. Try not to avoid making a lot of soap bubbles.

    The cell membranes are composed of a phospholipid bilayer. The shampoo or dishwasher soap helps dissolve the cell membrane.

    Sodium chloride helps remove histone proteins, which are bound to the DNA. Sodium chloride also neutralizes the charge repulsion that occurs between DNA strands. Sodium chloride helps keep the proteins dissolved in the aqueous layer so that they don't precipitate in the alcohol along the DNA.

    Papain breaks down the proteins down into polypeptides.

6. Make a filter cone and wet a paper filter with distilled water and place it in a test tube.

7. Pour the contents of the Ziploc bag into the filter and collect the liquid in a test tube. Collect about 3 mL liquid.

8. Add 2 volumes ice cold 95% ethanol to the strawberry liquid in the tube by pouring the ethanol carefully down the side of the tube so that it forms a separate layer on top of the strawberry liquid.

9. Watch for about a minute. You should see a white fluffy cloud at the interface between the two liquids. That's DNA!

10. Spin and stir the transfer pipet in the tangle of DNA, wrapping the DNA around the stirrer.

11. Remove the pipet and transfer the DNA to a piece of saran wrap or clean tube. The fibers are thousands and millions of DNA strands.

12. Rinse your funnel and test tube. Put the Ziploc bag in the trash.

Name: _____

OBSERVATIONS: ANSWER THE QUESTIONS ON THE DATA SHEET IN COMPLETE SENTENCES.

1. Where is DNA located in the cell?

2. What is the purpose of the soap in this activity?

3. What was the purpose of the sodium chloride? Include a discussion of polarity and charged particles.

4. What is the purpose of the enzyme papain in the lab exercise?

5. Why was the cold ethanol added to the soap and salt mixture?

6. Describe the appearance of your DNA.

7. Why do strawberries yield high amounts of DNA?

8. What steps in the procedure contributed to the release of the DNA from the strawberry plant cells?

9. What ingredients are in the extraction buffer?

10. Why is ethanol used for the precipitation or isolation of DNA?

20 LABORATORY

Introduction to Genetics

A. Instructional Objectives

After completing this exercise the student should be able to:
1. Explain Gregor Mendel's:
   a. principle of segregation using one pair of genes
   b. principle of independent assortment using two pairs of genes
2. Understand genetic terminology by using the following terms: gene, allele, phenotype, genotype, homozygous, heterozygous, self fertilize, and symbols for dominant and recessive genes.
3. Set up and work genetics problems using the following terms and concepts: monohybrid cross, P1, gametes, F1, F2, Punnett square, genotype, phenotype, genotypic and phenotypic ratios, test cross, and dihybrid cross.
4. Be able to determine genotypic and phenotypic ratios, and use data to deduce parental genotypes.

B. Introduction

Throughout human history there has been an awareness and concern with a wide variety of plants and animals affecting our well being. Long before there was an understanding of genetic mechanisms, specific matings of domestic organisms were made in an attempt to improve and increase the number of desirable traits in plants and animals. Organisms with desirable traits were mated and those lacking such traits were excluded from these matings. Humans were selecting for desirable traits through artificial selection as opposed to natural selection. By doing so, a wide variety of domestic animals have been produced from wild animals: dogs, cats, cattle, chickens, pigeons, horses, hogs, etc.

The genetic mechanisms involved in such matings were not explained until 1866 when Gregor Mendel reported his studies of

Adapted from *Laboratory Exercises for an Introduction to Biological Principles* by Gil Desha. Reprinted by permission of the author.

garden peas and stated what are now referred to as "Mendel's Principles." The significance of Mendel's work was not realized until shortly after 1900. Since that time tremendous knowledge of genetics has been accumulated. Beedle and Tatum have formulated a working model of gene action with their "one gene— one enzyme" theory. The genetic code and structure of DNA was explained by Watson and Crick in 1953. Jacob and Monod have theorized a mechanism for switching genes "on" and "off" in the operon theory.

Today an emphasis is being placed on human genetics and possible genetic manipulation or engineering. Conceivably such genetic manipulation could lead to the elimination or modification of deleterious genetic traits.

C. Materials

D. Methods

1. *Mendel's work and his principles:*

Gregor Mendel, an Austrian monk, was the first to make a systematic study of genetics and attempt to explain the mechanisms involved in the inheritance of physical traits. He published a scientific paper regarding his findings on the genetics of garden peas in 1866. This work explained how genes are involved in producing physical traits of pea plants and their seeds.

Genes are segments of DNA in specific locations on specific chromosomes that code for specific traits. Genes code for the particular amino acid sequence of every protein in living organisms. A "tall" gene in pea plants might produce more plant growth hormones, so the pea plant grows taller than other plants. A "short" gene might produce less plant growth hormones, leading to a shorter plant.

a. Mendel's Principle of Segregation: This principle states that each parent contributes only one allele of a gene to their progeny, or offspring, through that parent's gametes. An allele is an alternate form of a gene located at the same place on a pair of homologous chromosomes. Remember that most organisms are diploid and have 2 copies of each chromosome. Meiosis is the process responsible for segregating each member of the homologous pair into separate gametes.

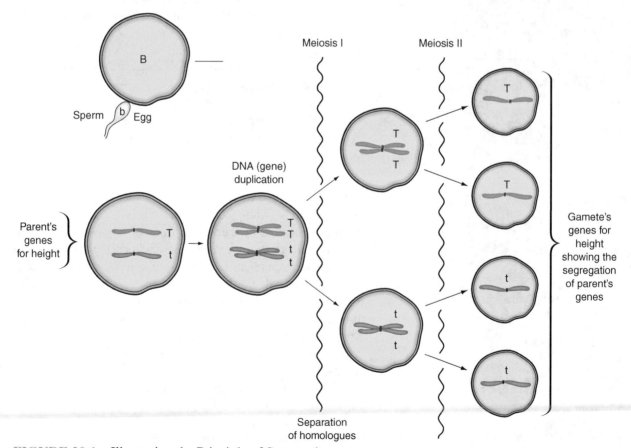

FIGURE 20.1   Illustrating the Principle of Segregation

Consider the example in Figure 20.1; T represents the allele for tallness in pea plants; t represents the allele for shortness. T and t are alleles since they are alternate forms of the same trait—the height of the plant. The parent plant has both alleles, T on one homologous chromosome and t on the other homologue. Recall that homologous chromosomes separate during meiosis. Only one of the two alleles from each parent is involved in producing the next generation; the alleles are segregated from each other.

You received one allele for each of your genes from your mother and one allele for each gene from your father. You will pass on one allele or the other to your offspring.

b. Mendel's **Principle of Independent Assortment**: This principle states that the segregation of one allelic pair or trait occurs independent of any other pair or trait. If you have two different genes on two separate chromosomes, each pair of homologous chromosomes behaves independent of other homologous chromosomes during meiosis. Remember that orientation of homologous pairs during metaphase I is a random event.

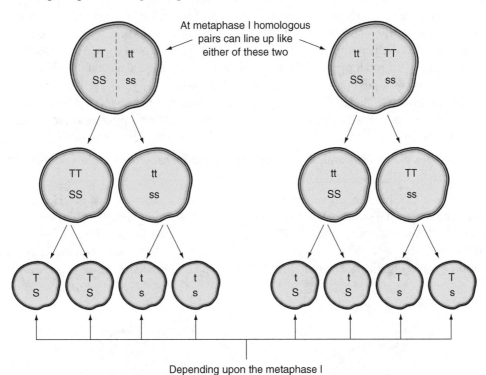

FIGURE 20.2 Principle of Independent Assortment

Consider the example in Figure 20.2; T is for tall, t is for short, S represents an allele for smooth seeds, and s represents an allele for a wrinkled seed coat. The parent plant has both alleles for both genes, Tt and Ss. In interphase, the chromosomes duplicate: TT, tt, SS, and ss. At metaphase I, TT and SS can line up on the left side, ultimately producing gametes of TS and ts. Or the TT and SS chromosome can line up on opposite sides of the dividing cell, producing gametes of tS and Ts. The principle of independent assortment says that you get all four possibilities, rather than keeping tall and smooth alleles together.

Note: in living organisms, many different genes are on the same chromosomes and the genes are said to be "linked" together, but we won't be considering those situations.

2. *Basic nomenclature:*
   a. **Monohybrid cross**—involves one gene for one trait. Keep in mind that most genes have at least 2 alleles for each trait. Crossing a tall pea plant to a short pea plant is an example of a monohybrid cross. While the single gene is plant height, it does have two alleles: tall and short. Other

genes may have more alleles. For example, your basic blood type is determined by combinations of three alleles: the A allele, the B allele, and the o allele.

b. **Dihybrid cross**—involves 2 genes responsible for 2 different traits. Crossing a tall pea plant with smooth seeds to a short pea plant with wrinkled seeds is an example of a dihybrid cross. The first gene is plant height; the second gene is seed appearance. Often a dihybrid cross involves 4 alleles, 2 for each gene.

c. **Phenotype**—refers to measurable traits and observable physical features of organisms. When referring to tall pea plants, tall is a phenotype because it can be measured and compared to short pea plants. Your hair color is a phenotype. If red in color, your phenotype for the hair color gene is red. If your eyes are brown, then your eye color phenotype is brown. Less obvious traits can also constitute phenotypes. Chances are that phenotypically you have type O blood that is also Rh positive. Your Rh type is a separate gene with two different alleles.

How many genes are responsible for type O positive blood? _____

d. **Genotype**—refers to the actual genetic makeup or formula for a specific phenotype. The genotype represents the combinations of alleles on the chromosomes. If you have brown eyes, you have inherited the genotype to produce the phenotype of brown eyes. Your genotype may be BB (two alleles for brown eyes) or Bb (one allele for brown eyes and one for blue eyes). Either of these genotypes would produce a brown-eyed phenotype. If you have blue eyes, your genotype would be bb (two alleles for blue eyes). Genotypes are expressed by letters of the alphabet, often corresponding to the first letter of dominant trait. Conventionally, the capital letters indicate dominant alleles and the lower case letters indicate the recessive allele.

Tall pea plants are dominant to short pea plants; what letter would you use to represent the short genotype? _____

e. **Homozygous**—has the same alleles. Homo = same, zygous = zygote. Homozygous literally refers to having the same 2 alleles in the zygote, or developing organism. Consider the example in Figure 20.3; if your genotype is BB, you have homozygous brown alleles for the eye color gene. If your genotype is bb, you also have homozygous alleles, in this case 2 blueness alleles. Homozygous organisms can only produce one type of gamete for the next generation.

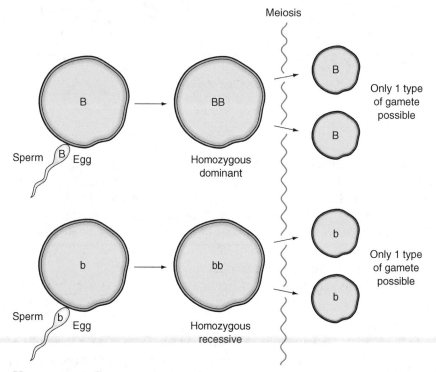

FIGURE 20.3   Homozygous Genotypes

f. **Heterozygous**—has different alleles. Hetero = different, zygous = zygote. If an organism has 2 different alleles for the same trait, it is said to be heterozygous. Consider the example in Figure 20.4; if your genotype is Bb, you are heterozygous for the eye color gene. Heterozygous individuals can produce 2 different types of gametes.

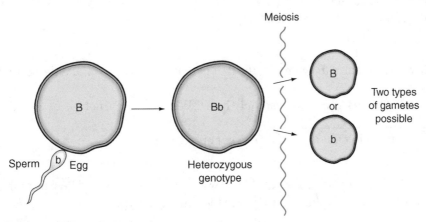

FIGURE 20.4   Heterozygous Genotypes

g. A **dominant** allele is one which will always produce its phenotype. Dominant alleles are indicated by a capital or upper case letter. The tall allele, T, is dominant over the short allele in plant height. A plant has to have at least one dominant allele in its genotype to have the tall phenotype. This T allele may be in combination with a second T. The genotype is then said to be homozygous dominant, TT. Both alleles are alike and can only produce one type of gamete, T.

h. However, the T may be in combination with a **recessive** allele which produces its phenotype only in the homozygous genotype. Recessive alleles are indicated by small letters corresponding to the letter of the dominant trait. (You would use a "t" to represent the short genotype rather than an "s" because the dominant allele is T for tall.) The recessive allele t may occur with the dominant T in the heterozygous genotype. The phenotype would be tall because the dominant allele will always produce the dominant phenotype.

Only a homozygous recessive genotype, tt, can produce a short plant phenotype. Keep in mind that a "homozygous recessive genotype" is redundant. A recessive genotype has to be homozygous to be recessive, so it is often referred to as simply a recessive genotype.
   What is the phenotype of a Tt pea plant? _____
   What is the genotype of a short pea plant? _____

i. In genetic crosses, **P1** refers to the parental generation. Two P1 individuals are crossed to produce the offspring or **F1** generation. F1 refers to the first filial generation. Some genetic crosses will continue to the **F2**, or second filial, generation.

3. *Genetic Problems (Monohybrid crosses):* If these simple rules are followed, little difficulty should be encountered in working genetics problems.
   a. Determine the P1 genotypes from the information provided. Write the genotype of one parent, follow this by an X to indicate the mating or crossing, and then list the other parent's genotype.
   b. Next list all possible gametes produced by each parent.
   c. Use a Punnett square to combine the gametes as would occur during fertilization. Write the gametes from the first parent on the top of Punnett square and fill in the boxes below each gamete. Write the gametes from the second parent down the left side of the Punnett square and fill in the boxes beside each gamete.
   d. Determine the genotype and phenotype of each potential individual produced.
   e. Determine the genotypic and phenotypic ratios, if the question asks, by counting how many individuals there are for each genotype and each phenotype.

Example:

A. Consider the dominant allele T for tall and the recessive allele t for short. Cross a homozygous tall plant with a homozygous short plant and determine the phenotype and genotype of the F1 generation.

The first step is to determine the genotypes of the parents. The genotype of the homozygous tall plant is TT. The genotype of the short plant is tt.

    a. P1   TT   ×   tt

    b. Gametes   T, T   t, t

    c. When you combine the gametes, the first parent can only give T and the second parent can only give t.

    d. The resulting F1 genotypes can only be Tt. What is the phenotype of an individual that is Tt?

    _____

    e. The phenotypic ratio is: all tall. The genotypic ratio is: all Tt.

Example:

Plants can **self fertilize**—the male part of a flower can fertilize the female part of the same flower. In genetic problems, an individual is crossed to the same genotype.

B. Allow previous F1 generation to self-fertilize. Determine the phenotypic and genotypic ratio of the F2 generation.

    a. The parents of this cross are _____ × _____

    b. The gametes are _____ _____

    c. The gametes from the first parent have been put along the top of the Punnett square in Figure 20.5. The gametes from the second parent are down the left side.

    d. Fill in the genotypes of the F2 individuals in the Punnett squares. Determine the phenotype of each individual and write it under the genotype in Figure 20.5.

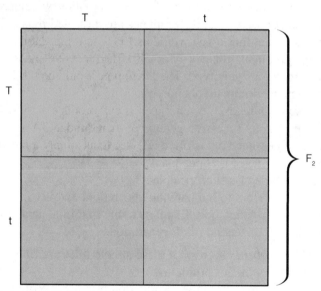

FIGURE 20.5   Determining the F$_2$ Generation

    e. What is the phenotypic ratio? _____

    f. What is the genotypic ratio? _____

C. In guinea pigs black fur is dominant over white fur. A black male guinea pig is mated with a black female guinea pig. In the course of 15 months, they have 27 black and 9 white offspring. The offspring illustrate what phenotypic ratio? Let's work through the problem step by step.

    What is the genotype of the white offspring? _____

    Where must each b have come from? _____

What are the genotypes of the parents? _____

What are the genotypes of the black offspring? _____

P1 _____ × _____

Gametes _____ × _____

What is the genotype of the white offspring? _____

What are the genotypes of the black offspring? _____

The offspring illustrate what phenotypic ratio? _____

D. A **test cross** crosses a homozygous recessive parent to a parent of an unknown genotype. Thus the unknown genotype can be determined.

Test cross a tall pea plant. What is the phenotype of the other plant? The "test cross" parent is always homozygous recessive; in this case a short plant.

What is the genotype of the short plant? _____

Is the unknown genotype dominant or recessive? Why? _____

_____

P1 T- (unknown genotype) x tt (the test cross parent, is always homozygous recessive)

   a. If the F1 offspring are all tall, the genotype of the tall parent is _____.

b. But if the F1 offspring are tall and short, the genotype of the tall parent is _____

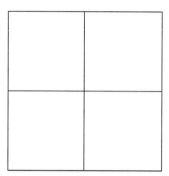

c. What is the value of a test cross?

4. *Probability:* One of the most useful facets of genetics problems is that they allow you to predict what the chances are for a particular phenotype occurring. Genetics is really a matter of **probability**, the likelihood of the occurrence of any particular outcome.

As a simple example, consider that the probability of coming up with heads in a single toss of a coin is one chance in two, or 1/2. Now apply this example to the question of the probability of having a certain genotype. For example, in the cross between two heterozygous tall plants (Tt), the genotype TT occurs in 1 of the 4 boxes of the Punnett squares. The probability of the genotype TT in the F1 generation is 1/4.

What is the probability of an individual, from the above problem, having a genotype of:

Tt _____

tt _____

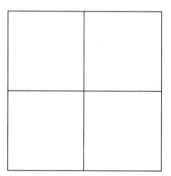

To extend this idea, consider the probability of flipping heads twice in a row with our coin. The chance of flipping heads the first time is $^1/_2$. The same is true for the second flip. The chance (probability) that we will flip heads twice in a row is not $^1/_2 + ^1/_2 = 1$. Instead of adding the probabilities, you have to multiply the probabilities: $1/2 \times 1/2 = 1/4$. The probability that we could flip heads three times in a row is $1/2 \times 1/2 \times 1/2 = 1/8$.

Returning to the previous genetic example, what is the probability that three offspring will each have the genotype TT? _____

What is the probability that three offspring will all be tall?

_____

While the probability of each child being tall is 3/4 or 75%, the probability of three children being tall is about 42%. Odds are one of the first 3 children will be short!

5. *Monohybrid, sex linked inheritance* In humans, as well as other primates, sex is determined by special sex chromosomes. An individual containing two X chromosomes is a female, while an individual possessing an X and a Y chromosome is a male.

What is your genotype? _____

What gametes are produced by a female? _____

_____

What gametes are produced by a male? _____

The gametes of which parent will determine the sex of the offspring? _____

The sex chromosomes bear alleles for traits, just like the other 22 pairs of chromosomes in our bodies. Genes that occur on the sex chromosomes are said to be sex linked. The Y chromosome is much smaller than its homologue, the X chromosome. Consequently, most of the genes present on the X chromosome are absent on the Y chromosome. In humans, color vision is a sex linked trait; the gene for color vision is located on the X chromosome but is absent from the Y chromosome. When working with sex linked traits, you do have to indicate the dominant (N) allele, the recessive (n) allele, the X chromosome, and the Y chromosome. X^N indicates a dominant trait on the X chromosome. X^n indicates a recessive trait on the X chromosome. Y indicates the gene is not present on the Y chromosome; it does not get an "N" or "n" because there is no allele present.

a. Normal color vision (X^N) is dominant over colorblindness (X^n). Suppose a colorblind man has children with a woman whose genotype is $X^N X^N$.

What is the genotype of the color blind father? _____

What proportion of the daughters would be colorblind? _____

What proportion of the sons would be colorblind? _____

b. One of the daughters from the previous question marries a color blind man.

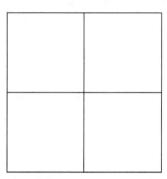

What proportion of their sons will be colorblind?

_____

Explain how a colorblind daughter might result from the above marriage.

_____

c. A woman who is heterozygous for the colorblind gene marries a man with normal vision. What is the genotype of the woman?

_____

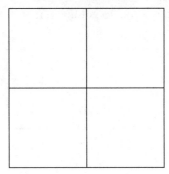

Assuming they have 4 children, what proportion of the children will be colorblind?

_____

What proportion of just the sons will be color-blind? _____

What proportion of just the daughters will be color-blind? _____

6. *Dihybrid crosses:* A dihybrid cross involves two separate traits where the alleles are located on two different pairs of homologous chromosomes. Dihybrid individuals form four distinct gametic types with theoretically equal frequencies because of random orientation during meiotic metaphase (Mendel's Law of Independent Assortment).

Example: In peas, tall is dominant over short and red is dominant to white. A heterozygous tall, red flower pea plant has a genotype of TtRr. It produces four different gametes: TR, Tr, tR, and tr. Notice that an allele (T or t) from the first trait (height) must be in combination with an allele (R or r) from the second gene (petal color).

Some students use the "foil" method to determine the gametic combinations. The "first" combination is T with R, T combines with the "outer" r, t combines with the "inner" R, and the "last" t combines with the last r.

Other students prefer a mathematical setup:

You could also use a Punnett square-like setup:

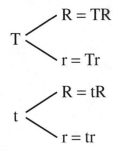

T — R = TR
— r = Tr

t — R = tR
— r = tr

Just be sure to realize that this is just to get the gametes; we aren't done yet!

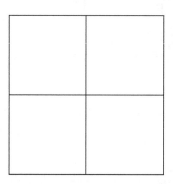

Often a heterozygous parent is crossed to another heterozygous parent, TtRr, which produces the same type of gametes. Because there are 4 possible gametes from the first parent and 4 possible gametes from the second parent, there are 16 possible genotypes in the offspring. You need to use a big Punnett square. Write the gametes of the first parent across the top of the Punnett square, and write the gametes of the second parent down the left side of the Punnett square, then fill in the offspring's genotypes, and determine the offspring's phenotypes.

P1   <u>TtRr</u>   ×   <u>   TtRr </u>

Gametes   <u>TR, Tr, tR, tr</u>   <u>TR, Tr, tR, tr</u>

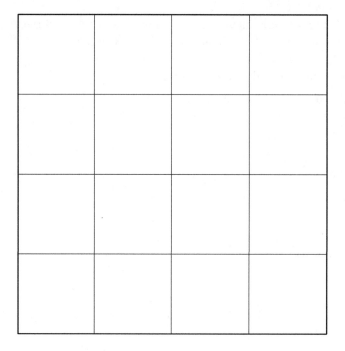

What are the 4 possible phenotypes?

_____

_____

_____

_____

What is the phenotypic ratio? _____

What is the genotypic ratio? _____

Name: _____

QUESTIONS

1. In peas the seed coat can be smooth or wrinkled. A wrinkled seed plant is crossed to a smooth seed plant. The F1 generation is all wrinkled. Which allele is recessive—smooth or wrinkled?

    Smooth

2. A wrinkled seed plant is test crossed. The F1 generation is all wrinkled. What is the genotype of the P1 wrinkled seed plant? What is the genotype of the F1 wrinkled seed plant?

3. Two wrinkled seed plants are crossed. The F1 generation has 80 wrinkled seed plants and 20 smooth seeds. What is the phenotypic ratio of the F1 generation? What is the genotype(s) of the P1 plants?

<table>
<tr><td></td><td></td></tr>
<tr><td></td><td></td></tr>
</table>

4. Right handedness is dominant to left handedness. If you cross 2 heterozygous individuals, what is the probability of getting a left handed child?
   What is the probability of getting 2 right handed children?

<table>
<tr><td></td><td></td></tr>
<tr><td></td><td></td></tr>
</table>

5. Green pods and wrinkled seeds are dominant genes. A pea plant that is homozygous for green pods and homozygous for wrinkled seeds is crossed to a yellow pod, smooth seed plant. The resulting offspring are self fertilized. What is the phenotypic ratio of the F2 generation?
   What is the probability of getting a plant that has a yellow pod and smooth seeds?

<table>
<tr><td></td><td></td><td></td><td></td></tr>
<tr><td></td><td></td><td></td><td></td></tr>
<tr><td></td><td></td><td></td><td></td></tr>
<tr><td></td><td></td><td></td><td></td></tr>
</table>

6. (Optional) Green pods, wrinkled seeds, and tall are dominant genes. A pea plant that is heterozygous for green pods, homozygous for smooth seeds, and homozygous tall is crossed to a pea plant that is heterozygous for green pods, homozygous for wrinkled seeds, and homozygous short. What is the phenotypic ratio of the offspring? (Hint: think about what gametic combinations each parent can produce; how big of a Punnett square do you really need?)

What are the genotypes of the parents?_____ x _____

What gametes can each parent give? _____ _____

<table>
<tr><td></td><td></td></tr>
<tr><td></td><td></td></tr>
</table>

21 LABORATORY

Human Genetics

A. Instructional Objectives

After completing this exercise the student should be able to

1. Determine his or her phenotype (and in some cases the genotype) for the genetically influenced traits listed in the exercise.
2. Given the genotypes for both the A, B, O, and Rh factors of the parents, determine the possible phenotypes of their offspring.
3. Shown a pedigree analysis chart, determine if the trait is dominant, recessive, or sex linked, and determine the genotypes and phenotypes of selected individuals in different generations.
4. Define the following terms: albinism, Huntington's Chorea, hemophilia

B. Introduction

Of all the subdivisions in biology many students find genetics to be the most intriguing. Everyone has some interest in why he is like he is; why her hair, eyes, and skin are certain colors; or why he is a human rather than a dog, a chicken, or an amoeba. The answers to these and related questions constitute the science of genetics.

Humans are not the best form of life for genetic study. Much more is known about inheritance in peas, corn, and fruit flies than is known about inheritance in humans. The basic principles of genetics were determined through the study of garden peas. Corn has been studied so extensively that breeders know exactly which four sets of grandparents to interbreed to get the best possible hybrid plant. Flies are bred by the billions, studied, and counted in laboratories around the world. We humans are less well understood than other organisms because we are tremendously complex, because we reproduce slowly, and because society frowns on human breeding experiments.

Adopted from *Laboratory Exercises for an Introduction to Biological Principles* by Gil Desha. Reprinted by permission of the author.

In spite of the obstacles and because of student interest, we will direct our efforts toward the study of humans rather than of peas, corn, or flies. Constructing a family tree is a time-honored method of studying human genetics. Particular traits may be followed through generation after generation. Sometimes a trait seems to "disappear" for awhile only to reappear in a great-great grandchild. The study of how such traits are inherited is aided by pedigree analysis. By studying family pedigrees we may be able to decide whether a trait is dominant or recessive in inheritance.

In this exercise we will determine some of the traits present in ourselves and study how these traits may have been inherited.

C. Materials

P.T.C. paper

D. Methods

In this study each student will determine their own phenotype and possible genotype in regards to several rather superficial traits. As you remember, the phenotype is the expression of the characteristic while the genotype is the actual genetic makeup of the characteristic. For example, you may possess ear lobes which hang free rather than attached ear lobes; free ear lobes would be your phenotype. Let "L" represent the dominant gene for free ear lobes and "l" for the recessive phenotype.

Sometimes you can determine your exact genotype. If someone has the attached ear lobe phenotype, their genotype is known and written as ll. But what if you have the dominant free ear lobe? Your genotype may be one of two possibilities: LL or Ll.

If one of your parents has the recessive attached ear lobe trait, what would your genotype have to be? _____

But what if both your parents also have the dominant free ear lobe phenotype? You would be unable to determine your exact genotype. Where doubt exists as to the exact genotype, you would indicate the unknown allele with a dash (-). If you and both your parents have free ear lobes, all three genotypes would be written as L-.

You are to determine your phenotype and genotype (as much as it can be determined) for each trait listed below. Record this data in the Results section.

1. Ear lobes. As mentioned, the free ear lobe is apparently dominant over attached ear lobes. Show the presence of free ear lobes as L- and attached lobes as ll.

2. Freckles. Having small spots of dark pigment, especially on the face and arms, is a dominant trait called freckles. Use F- for freckles and ff for no pigment spots.

3. PTC taste. Some individuals can taste a chemical substance designated as PTC (phenylthiourea) and others cannot. Put a small piece of PTC paper on your tongue; you will know if you taste it! Tasters possess dominant P allele, while non-tasters are homozygous recessive, pp.

   There are other papers available to taste as well; tasting other papers is still dominant. But the genes are on different chromosomes, so the genes assort independently—just because you taste the PTC paper, doesn't mean you will taste the other papers.

4. Hairline. In some people the hairline drops downward and forms a distinct point in the center of the forehead. This is known as a widow's peak. It results from the action of a dominant allele (W). A continuous hairline results from the recessive genotype (ww). Determine your phenotype by examining your hairline for a widow's peak or a continuous hairline. (You may have to skip this question if the gene for baldness has had its effect at the front of your head.)

5. Writing hand. Although environment can play a part, a dominant allele seems to cause many people to write predominantly with their right hand, R- genotype. Left handedness is caused by a recessive genotype. The genetics of ambidextrous, using both hands equally, is currently unknown.

6. Hitchhiker's thumb. This characteristic, known in anatomical terms as "distal hyperextensibility of the thumb," can be determined by bending the distal joint of the thumb back as far as possible. While there is some degree of continuous variation, some people can bend their thumb back until there is

almost a 90 degree angle between the two bones. Scientific evidence indicates that this is due to a recessive allele (h). There is some variation in gene expression; occasionally it will be found in only one thumb. In addition, it should be mentioned that there seems to be a 5% reduction in penetrance—that is, about one person in 20 who carries the recessive allele will not express the characteristic.

7. **Mid-digital hair**. Some people have hair on the second or middle phalange of one or more of their fingers, while others do not. The complete absence of hair from all fingers is due to a recessive allele (m) and the presence of any hair is due to a dominant allele (M). There also seem to be a number of different alleles which determine whether the hair grows on one, two, three, or all four fingers. This hair may be very fine. You should use a hand lens and look very carefully on all fingers before deciding whether this hair is completely absent from all of your fingers. The presence of even one hair on one finger gives you a genotype of M-.

8. **Interlocking fingers**. When the fingers are interlocked, some people will almost invariably place the left thumb on top of the right and others will place the right thumb over the left. Studies of family pedigrees indicate that the placing of the left thumb over the right is due to a dominant allele (T), while the right thumb on top is due to a recessive allele (t).

A, B, O blood types. There are three alleles responsible for the A, B, O, and AB blood types. A and B alleles are of equal dominance. The o allele is a recessive to both the A allele and the B allele. If your blood type is type A, your genotype could be AA or Ao. If your phenotype is type B blood, your genotype could be BB or Bo. Type AB blood comes from an AB genotype. And the homozygous recessive genotype of oo produces type O blood.

Can two parents with type B blood produce a child with type O blood? _____

Rh factor. Since the original study in which blood was grouped into the four types above, other blood factors have since been discovered. One such item, which has important medical implications, is the Rh factor. Most individuals have this factor and are said to be Rh positive. The presence of the Rh factor is governed by a dominant allele "R." The recessive allele is "r." When recording the genotype for the Rh factor, show Rh positive blood as R- and Rh negative blood as rr.

Can 2 Rh negative parents produce an Rh positive child? _____

Pedigree analysis. In these examples, black circles represent females with a particular trait, while clear circles indicate females who do not express the trait. Shaded squares indicate a male with the trait and white squares represent a male without the trait in question. Roman numerals (I, II) indicate generations and Arabic numbers (1, 2) indicate individuals. Examine the following pedigree in Figure 21.1. We may refer to a particular individual, IV-2, as being a male with a particular trait or II-5 as a female without the trait. All pedigrees have individuals "affected" with some trait; geneticists analyze patterns to determine whether the trait is inherited in a dominant, recessive, or sex linked pattern.

In this pedigree, each individual who expresses the trait has a parent who also expressed the trait. This pedigree illustrates the inheritance of a dominant trait. A trait determined by a **dominant allele** will not appear in an offspring unless it also appears in at least one parent. No generations are skipped. The dominant trait often affects about half of the children if even one parent is affected.

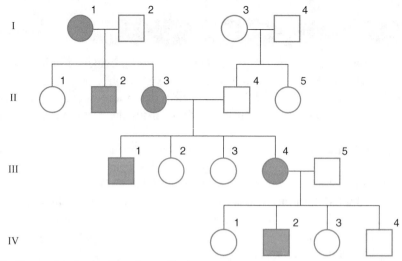

FIGURE 21.1   Pedigree Analysis, Dominant Trait

Recessive traits frequently skip one or more generations in a direct ancestral line. But if both parents have the trait, all of the children should be affected. In Figure 21.2 we are dealing with a recessive allele. Note that parents II-2 and II-3 do not express the trait, but their child III-2 does.

What are the genotypes of II-2 and II-3? _____

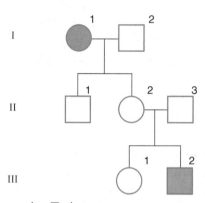

FIGURE 21.2   Pedigree Analysis, Recessive Trait

A **sex linked trait** affects more males than females. If you cross a carrier female to a normal male, neither parent is affected, none of the daughters will be affected, but half of the sons will be affected. If a female is affected, all of her sons will be affected. Because most sex linked traits are also recessive, they frequently skip a generation. Often a grandfather is affected, his sons and daughters are not affected, but his grandsons are affected. See Figure 21.3.

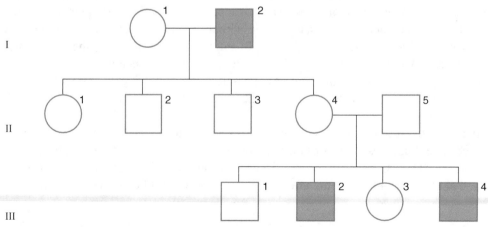

FIGURE 21.3   Pedigree Analysis, Sex Linked Trait

Name: _____

QUESTIONS

1. Record you data.

Hereditary Traits	Your Phenotype	Your Genotype
Earlobe attachment	_____	_____
Freckles	_____	_____
PTC tasting	_____	_____
Hairline	_____	_____
Writing hand	_____	_____
Hitchhiker's thumb	_____	_____
Mid-digital hair	_____	_____
Interlocking fingers	_____	_____

2. Finish the dihybrid problem below and indicate all the possible phenotypes of the offspring.

AB Rr × Bo Rr

FIGURE 21.4   Determining Blood Types

What are the possible phenotypes?

_____

_____

_____

_____

3. In humans, **albinism** (the lack of pigmentation) is an inherited trait. In the following pedigree individuals with albinism are indicated in black.

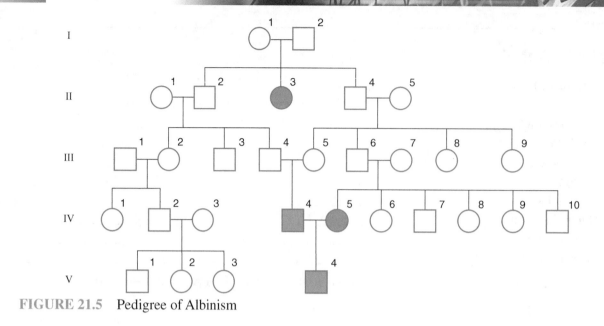

FIGURE 21.5   Pedigree of Albinism

a. Is albinism passed as a dominant, recessive, or sex linked trait?

b. What is the genotype of individual III-4 if we use "A" for the dominant trait and "a" for the recessive trait?

c. What is the genotype of IV-3?

d. Are individuals III-4 and III-5 related?

e. What are the genotypes of II-2 and II-4?

f. Given that I-1 and I-2 are heterozygous and cousins III-4 and III-5 marry, what is the probability of IV-4 having albinism? (Hint: multiply the probabilities of each mating producing the various genotypes.)

Probability of II-2 being heterozygous: _____

Probability of II-4 being heterozygous: _____

Probability of III-4 being heterozygous: _____

Probability of III-5 being heterozygous: _____

Probability of IV-4 being homozygous: _____

_____ x _____ x _____ x _____ x _____ = _____

The probability of 2 random (non-related) individuals having a child with albinism is 1 in 10,000 or about .01%. The probability of 2 cousins having a child with any recessive trait is about 100 times greater.

4. **Huntington's Chorea** is a mental disorder in which mental deterioration is accompanied by uncontrollable, involuntary muscular movements. At the present time, it is always fatal. Examine the following pedigree.

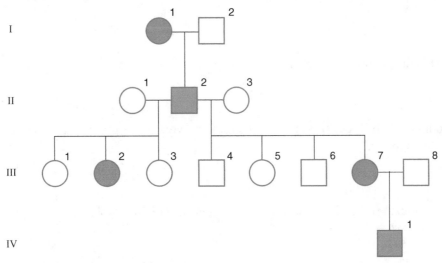

FIGURE 21.6   Pedigree of Huntington's Chorea

   a. Is Huntington's Chorea passed as a dominant, recessive, or sex linked trait?

   b. What is the genotype (use H or h) of individual II-1?

   c. Are individuals II-1 and II-3 related?

5. People with **hemophilia** are missing one or more blood clotting factors which make it difficult to impossible for their blood to clot. Most hemophiliacs are at risk of bleeding to death. Examine the following simplified pedigree of selected royal families in Europe.

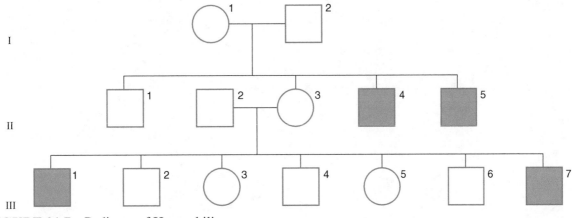

FIGURE 21.7   Pedigree of Hemophilia

a. Is hemophilia passed as a dominant, recessive, or sex linked trait?

b. What is the genotype of individual II-2?

c. What is the genotype of individual II-3?

d. If individuals II-2 and II-3 have more children, what is the probability their next daughter will have hemophilia?

e. If individuals II-2 and II-3 have more children, what is the probability their next son will have hemophilia?

f. If individuals II-2 and II-3 have more children what is the probability then next child will have hemophilia?

22

L A B O R A T O R Y

A Model of Epidemiology

Part A. A Model of Epidemic

A. Instructional Objectives: After completing this exercise the student should be able to:
1. Discuss what constitutes a communicable (infectious) disease.
2. Discuss the difference between an epidemic and a pandemic.
3. Using the model of this laboratory exercise, show how a communicable disease may be spread.

B. Materials

24 test tubes numbered 1–24   Saturated solution NaCl

24 Beral-type pipettes   1 dropping bottle 0.1 M AgNO$_3$

Test tube racks

Video: *Evolutionary Arms Race*

C. Introduction

The World Health Organization, commonly known as WHO (www.who.int/topics/infectious_diseases/en/), defines an infectious disease, also called a communicable disease, as a disease caused by a pathogenic microorganism—a bacterium, virus, parasite, or fungus. Pathogens are defined as parasites that cause disease. Infectious diseases can be spread directly or indirectly from one individual to another; that is, they can be spread from person to person. Sexually transmitted infections (STIs) are an example of person-to-person transmission. These diseases are spread by sexual activity when an infected individual comes in contact with an uninfected person. Common STIs are either bacterial or viral. Gonorrhea, syphilis, and chlamydia are examples of common bacterial STIs. The good news is that most bacterial STIs can be treated with antibiotics. The bad news is that there are antibiotic-resistant strains of syphilis and gonorrhea. More bad news is that viral STIs cannot be cured, but they can be treated. Once infected, the individual has a viral STI for life. Common viral STIs are herpes simplex (HSV) type II and human papillomavirus (HPV). Cervical and penile cancers can be the result of certain strains HPV infection. Human immunodeficiency virus (HIV) is a deadly STI that is ravaging parts of Africa. In Africa, HIV is transmitted as an STI. In the United States HIV is transmitted primarily as MSM (men having sex with men) and intravenous (IV) drug use. There is a new trend in HIV infections in the United States: increasing numbers of infections by WSM (women having sex with men). Of this population of women, the women who are particularly being infected are those who African American.

Indirect transmission of pathogens can occur through contaminated water, food, animals, and inanimate objects. Although in the United States waterborne pathogens are a rarity, there are parts of the world where water quality is nonexistent. Diarrheal diseases resulting from exposure to *Escherichia coli* and *Vibrio cholera* kill 2 million children younger than age five each year. Outbreaks of cholera, a waterborne disease, are common in developing countries when there is poverty, armed conflict, environmental catastrophes, or poor sanitation of water supplies. In areas of poor sanitation, people use the same water source for sewage disposal and as a source of water for drinking, washing, and bathing purposes. Food may become contaminated with pathogens. In the spring of 2007, organic spinach in California became contaminated with *E. coli*, a common bacterium found in the colon of mammals. Some people who ate the raw spinach became ill with food poisoning. In 1993, some children in Seattle, Washington, ate partially rare hamburgers at a Jack in the Box fast-food restaurant. The children died from becoming ill with a strain of bovine *E. coli*. The bacterium caused kidney failure. Eggs, processed pork, and chicken can be sources of salmonella.

In the *Evolution* series on PBS, in the episode titled the *Evolutionary Arms Race* (www.pbs.org/wgbh/evolution/library/10/4/l_104_09.html), the scene is presented where a number of prisoners are being kept in cramped areas with little or no air circulation. There is a definite visual account of overcrowding. As a result of overcrowding, the prisoners may become infected by inhaling the droplets of fluids spread into the air by coughing and sneezing. Drug-resistant tuberculosis is an epidemic in Russian prisons. Since the fall of communism in Russia, the number of incarcerated has dramatically risen and prison overcrowding is very common. When a person coughs, the microscopic droplets can float in the air for many hours. Pathogens can be found on tablecloths, table wear, glassware, and door handles. One simple way to slow the spread of disease is to wash your hands periodically.

The study of how diseases are spread is called epidemiology. An epidemiologist is a medical detective who studies how, when, where, what, and who are involved in an outbreak of a disease. An epidemiologist makes a comparison of the incidence of new cases of a specific disease to the previous incidences of cases of that disease. If there is a significant increase in a given time period within a specific geographic area, then an epidemic may be in progress. If the outbreak covers two or more continents, then the more appropriate term is a *pandemic*. With intercontinental air traffic being affordable, the possibility of a worldwide pandemic becomes increasingly possible.

In this part of the lab, you will discover the pathway of infection of a "virus." The test tubes will represent the bodily secretions of an individual. One person will be infected and will be the carrier of the "virus." There will be three exchanges of bodily fluids.

D. Procedure
  1. Transmission of the "virus."
    a. Rules for the simulation:
      (1) Each student will receive a numbered test tube half-filled with a liquid. The liquid represents bodily secretion. The test tubes contain distilled water except for one tube containing a "virus."
      (2) The "virus" is spread by direct body contact.
      (3) Record the letter of the test tube you obtained.
    b. Exchange 1—Find one lab member at random and exchange solutions by giving each other a Beral pipette full of the solution in the test tube.

c. Record the letter of your possible exposure to the "virus" on Table 22.1.
d. Exchange 2—At random, find a different lab mate and repeat the exchange of solutions by giving each other a Beral pipette full of the solution in the test tube.
e. Record the name of your possible exposure to the "virus" on Table 22.1.
f. Exchange 3—At random, find a different lab mate and repeat the exchange of solutions by giving each other a Beral pipette full of the solution in the test tube.
g. Record the number of your possible exposure to the "virus" on Table 22.1.

2. Identification of the carrier
   a. Your lab instructor will test all of the lab mates with an indicator solution.
   b. If you have become infected with the "virus," the liquid in the test tube will become a cloudy white.
   c. If you have become infected, write your letter on the board and the sequence of your personal contacts.
   d. Record the class results on Table 22.2.
   e. Draw a diagram in Table 22.3 that shows the sequence of transmission through the three exchanges.
   f. Your lab instructor will identify the letter of original carrier of the virus.

TABLE 22.1   A Personal Contacts

Exchange	One	Two	Three
Contact Letter			

TABLE 22.2   **A** Discovery of the Pathway of Transmission

Student	Contact One	Contact Two	Contact Three

TABLE 22.3 A Route of Transmission of the "Virus"

Name: _____

QUESTIONS

1. What is the maximum number of students who are "infected" with the "virus" before the simulations begins? _____

2. What is the maximum number of students who are "infected" with the "virus" after the first exchange? _____

3. What is the maximum number of students who are "infected" with the "virus" after the second exchange, assuming no one met with the same person in the previous exchange? _____

4. What is the maximum number of students who are "infected" with the "virus" after the third exchange, assuming no one met with the same person in any previous exchanges? _____

5. If the rounds were extended to Round 4, what is the maximum number of students who would be "infected" with the "virus"? _____

6. Why might the observed number of persons infected by the "virus" be lower than the maximum possible?

7. If one of your lab mates had a highly communicable viral disease, how likely would it be that you could be exposed to the disease?

8. Suppose that you could have multiple contacts during the time period for an exposure. What differences might be seen in the spread of the disease?

9. What are some ways that infectious disease may be transmitted from one person to another?

10. What are some ways to prevent the transmission of infectious disease from one person to another?

Part B. Parasitic Infections

Instructional Objectives
1. Be able to distinguish between the following terms:
   a. definite host and intermediate host.
   b. ectoparasite and endoparasite.
   c. prevention and control.
   d. predation and parasitism.
   e. parasite and host.

2. Be able to identify the following parasites from microscopic slides and specimens:
   a. *Plasmodium* (malarial protozoan).
   b. *Dermacentor* (tick) .
   c. *Taenia* (tapeworm).
   d. *Enterobius* (pin worm).
   e. *Ascaris* (round worm).
   f. *Clonorchis* (liver fluke).
   g. *Trichenella* (trichina worm).
   h. *Schistosoma* .

3. Be able to explain the role of vectors in the transmission of disease.

4. Be able to identify the stages of the life cycles of the parasites in this laboratory exercise.

Materials

Microscopes
Dissecting microscopes
Microscopic slides of the following:
   a. *Plasmodium* (malarial protozoan)
   b. *Dermacentor* (tick)
   c. *Taenia* (tapeworm)
   d. *Enterobius* (pin worm)
   e. *Clonorchis* (liver fluke)
   f. *Trichenella* (trichina worm)
   g. *Schistosoma*
   h. *Dermacentor* (tick)

specimens of the following:
Dermacentor (tick)
Ascaris

Introduction

Parasitology is the study of parasites, their hosts, and the relationship between them. Ecologically, parasitism is defined as a form of predation where the predator is an organism that feeds on another organism, its prey. Usually the predators are larger than their prey. However, in this special form of predation known as parasitism the predator is smaller than its prey.

Parasitism is also defined as a type of symbiotic relationship. In this symbiotic relationship, one organism (the parasite) benefits from the relationship while the other organism (the host) does not. The parasite is "happy" while it makes its host "miserable." Just think about a dog that has a bad case of fleas. The fleas are happy biting it and sucking the blood from the dog while the dog is miserable trying to scratch and bite the fleas in order to kill them. The parasite depends upon the host for its food source and residence. The parasite does not immediately kill its host but it can definitely weaken the host.

Parasites may be viruses, certain species of bacteria, certain species of fungi, and invertebrates. Invertebrate parasites are usually worms (flatworms and roundworms) and arachnids (lice and ticks). There is even a vertebrate parasite called the lamprey. The lamprey is a jawless fish that attaches to the side of a larger fish and by using its rasping tongue forms a hole into the body cavity. It then sucks the blood and body fluids of the larger fish. Parasites may be transmitted by insects or arachnid carriers called vectors.

There are two types of parasites: ectoparasites and endoparasites. An ectoparasite resides on the surface its host. Ticks and lice are two types of ectoparasites that live on warm-blooded organisms. Both of these arachnids suck on the blood of their hosts. Endoparasites live within the body of their host by invading internal organs or attaching to the ducts of internal organs such as the digestive and respiratory systems or in the blood or brain. Examples of endoparasites are tapeworms, flukes, and roundworms.

Parasites may have a simple or complicated life cycle. Those that have a complex life cycle may have one or more intermediate hosts. The intermediate host houses the sexually immature stage of the parasite life cycle. The parasite reaches sexual maturity in its definite host.

Procedure

1. Plasmodium

Plasmodium, a sporozoan, is the pathogen responsible for the disease malaria. The name malaria literally means "bad air." It was thought that if a person breathed in the night air, that person could become ill with malaria. Malaria is one of the greatest scourges of mankind. Each year 350 to 500 million cases of malaria occur worldwide. This disease is particularly widespread in sub-Saharan Africa, where over 90% of malaria-related deaths occur. Over one million people die, most of them young children under the age of five. Almost two-thirds of all cases of malaria-related deaths occur among the poorest 20% of the world's population. Individuals in these areas do not have access to simple preventative measures like mosquito netting, medicines, proper drainage of stagnant water, and pesticides. Most of the people become infected with malaria through the bite of an infected female Anopheles mosquito. The female needs blood to fertilize her eggs. A handful may become infected through blood transfusions, organ transplants, or IV drug use. The symptoms of malaria include the following: severe chills, fever, sweating, an enlarged and tender spleen, confusion, and great thirst. People succumb to this disease due to anemia, kidney failure, or brain damage.

There are four different species of Plasmodium that infect humans. Table 23.1 lists characteristics of the four species.

The life cycle of the Plasmodium is complicated. There are two hosts required for the life cycle: the female Anopheles mosquito (the intermediate host) and the human (definite host). The female Anopheles mosquito is a vector and is not affected by the parasite. Following are the events of the life cycle of Plasmodium:

1. An infected female Anopheles mosquito bites the human host and injects sporozoites into the bloodstream.
2. Sporozoites invade the liver, where they develop into merozoites.
3. The merozoites invade red blood cells, where they reproduce.
4. Periodically, merizoites break out of the red cells, bringing on the chills and fever characteristic of malaria.
5. Eventually some merozoites develop into either male or female gametocytes.
6. These will die unless they are sucked up by the bite of a female Anopheles mosquito.

TABLE 23.B.1   Characteristics of the Four Species of Plasmodium

Species	Location	Incubation	Characteristics
P. vivax	India and Central and South America	8–13 days	Sometimes leads to life-threatening rupture of the spleen
P. ovale	Africa	8–17 days	Can hide in the liver of partially treated patient and return later
P. malariae	Worldwide	2–4 weeks	If untreated, may last many years
P. falciparum	Worldwide	5–12 days	Most life-threatening form of malaria; resistant to major antimalarial drugs

7. Once in the stomach of the mosquito, the gametocytes form gametes: sperm and eggs.

8. Sperm and eggs fuse to form zygotes.

9. The zygote invades the stomach wall of the mosquito, forming thousands of sporozoites.

10. Sporozoites migrate to the salivary gland, ready to be injected into a new human host.
Examine the blood smear slide containing the malaria parasite.

2. Tapeworm

Taenia (tapeworm) is a flatworm. Tapeworms live exclusively as parasites within the digestive tracts of other animals. A mature tapeworm's body is ribbon-like in shape. The head of the tapeworm is the scolex, which has hooks and suckers to hang on the intestinal wall of the host. The scolex asexually produces a series of proglottids. Tapeworms neither have a digestive system nor do they secrete digestive enzymes; they absorb digested food formed in the intestine of their host. The immature proglottids are nearest to the scolex, while the sexually mature proglottids are the furthest from the scolex. Tapeworms are true hermaphrodites. Each proglottid contains both male and female reproductive organs. A chain of proglottids is called a strobilus. Some strobili may reach a length of 15 to 30 feet. Mature proglottids are filled with eggs and are shed in the feces. The proglottids may crawl out of the feces onto vegetation, where they may be eaten by an intermediate host.

The eggs are eaten by an intermediate host such as hogs, fish, and cattle. Eggs enter the circulatory duct where they are carried into the intermediate host's muscle. Once in the muscle, a cysticercus larvae is formed. A cysticercus is a bladder-like structure containing a scolex. The cysticercus everts its scolex and attaches to wall of the intestine. People infected with tapeworms are usually asymptomatic. However, abdominal pain, diarrhea, and weight loss may occur. In the Victorian era, women used tapeworms as a means of weight loss. The disease is diagnosed by finding proglottids or eggs in the feces. Prevention of tapeworms involves the thorough cooking of beef, pork, and fish. The beef tapeworm is *Taenia saginata*, the pork tapeworm is *Taenia solium,* and the fish tapeworm is *Dipyllobithrium latum*.

Examine the prepared slide under scanning or low power.

3. Flukes

Flukes lack hooks as do tapeworms, but they do have suckers for attaching to the internal body cavities of their hosts. Some flukes have a digestive system. They may live in the respiratory tract, digestive tract, the blood, or in the brain. Flukes have a complicated life cycle. *Clonorchis sinensis* live in the bile ducts of humans, cats, dogs, and pigs. It is very common in Asia. The length of the worm is 1 to 2 cm. Although hermaphrodites, they cross-fertilize. Following are the events of the life cycle of the human liver fluke:

1. Eggs are shed in the feces. Eggs contain a complete, ciliated larva called a miracidium.

2. Miracidium are eaten by snails, the first intermediate host. Within the snail, the miracidium give rise to a free-living cercariae.

3. The cercaria leave the snail and burrow into a fish, the second intermediate host. In the fish the fluke encysts as a metacercaiae.

4. When ingested, metacercariae become juvenile flukes that migrate from the gut into bile ducts.

Light infections are usually asymptomatic. But, heavy infestations may cause bile duct obstructions and serious liver disease. Thorough cooking of fish destroys the metacercariae.

Examine the prepared slide using scanning or low power only.

4. Schistosoma

One of the most important of the blood flukes is genus *Schistosoma*. They affect 1 in 20 of the world's population. More than 200 million people in tropical Asia, Africa, Latin America, and the Middle East are infected. 800,000 each year die from the disease *schistosomaisis*. The blood fluke can damage the liver, lungs, intestines, and urinary bladder. Proper sanitation is important in preventing the spread of the worm or lowering the population of snails in the water.

Following are the events of the life cycle *of Schistosoma*:

1. Eggs leave the body through the urine and the feces.
2. If the eggs reach water, they hatch into miracidia.
3. Miracidia must reach a freshwater snail within hours of hatching.
4. Within the snail, miracidia develop into sporocysts.
5. Sporocysts develop into cercariae.
6. Cercariae attack human skin.
7. Cercariae enter the circulatory system and are swept to the lungs.
8. Cercariae renter the bloodstream and migrate to the hepatic and portal blood vessels.
9. Worms sexually mature and pair on their way to the upper intestine, lower intestine, or bladder.
10. Ulceration of the intestinal and bladder wall occurs. Cirrhosis of the liver occurs due to blockage of the liver capillaries.
    Examine the prepared slide under low power.

5. Ascaris

Ascaris, a round worm inhabiting the small intestine, is the causatic agent of *acariasis*. Females reach a length of 10 to 12 inches. The males are considerably smaller. The crook in the posterior end of the male's body is an adaptation for copulating with the female. Infection of this worm occurs in tropical areas where sanitation and personal hygiene are poor. Infection occurs when eggs are ingested. The hatched eggs liberate larvae that migrate from the small intestine through the lymphatic system or blood stream to the alveoli of the lungs. From the alveoli the larvae ascend the respiratory tract into the pharynx, where they are swallowed. Larval migration through the lungs produces fever, cough, wheezing, and other associated respiratory problems. The adult worms may cause absorption problems that result in weight loss, abdominal cramping, and rarely, intestinal obstruction. Female worms produce thousands of eggs which appear in fecal smears. Occasionally adult worms are vomited or found in a stool. Examine the specimen.

6. Trichinella

Trichinella, another roundworm, causes *trichinosis*. The disease is commonly contracted by eating improperly cooked pork or pork products such as sausage. There have been incidences where hunters have eaten improperly cooked bear meat and come down with trichinosis. Pork should be cooked to a temperature of 170°F. Infected meat contains *calcareous cysts* filled with larvae.

Digestion of the cyst by digestive enzymes releases the larvae. The larvae penetrate the intestinal wall. The larvae immediately mature and mate. After mating the females burrow deeper into the gut's mucosa, where they liberate larvae directly into the lymphatic system and blood vessels. The larvae are carried to skeletal muscles and invade individual muscle fibers of the eyeball, diaphragm, and tongue. Encysted larvae can remain there for years. Adult worms may cause abdominal cramping. More serious symptoms are muscle soreness, pain, fever, chills, and prostration. These symptoms are much like the flu. Other, more serious aspects may occur if the heart and central nervous system are invaded by the parasite. Symptoms usually disappear after three months. Diagnosis involves blood tests. An elevated eosinophil count is an indicator of worm infestation. Steroids may be needed for severe disease complications involving the heart or allergic reactions. Control involves feeding hogs garbage that has been thoroughly cooked. All pork products need to be thoroughly cooked. Freezing pork to −15°C for three weeks or −18°C for one day kills encysted larvae.

Examine the microscope slide showing encysted larvae.

7. Enterobius

Enterobius, the pinworm, is another example of a parasitic roundworm. This common pest affects 20% of children and up to 90% of institutionalized children. Adult pinworms are small when compared to *Ascaris*. Females are from 8 to 13 mm in length, while males are 2 to 5 mm long. Males have a curved tail, a modification for copulation. Adult pinworms inhabit the colon. Female worms migrate through the anus at night and deposit eggs in the perianal area (the area around the anus). Up to 11,000 eggs are deposited in the skin folds. The female's movement and the gelatinous substance encasing the eggs cause an intense itching. Under ideal conditions eggs can survive up to three weeks on clothing, bedding, and toys. Eggs are transferred from contaminated articles (fomites) or fingers to the mouth and are swallowed. Airborne eggs may be inhaled or swallowed. Once in the small intestine, the eggs hatch. The young mature and feed on the contents within the colon. Occasionally the eggs hatch perianally and enter the colon through the anus. Pin worms are not a serious health threat. Diagnosis involves finding eggs periannally or samples taken on a piece of Scotch tape. Treatment involves medicating the entire family and careful cleaning of the household to rid fomites of eggs. If this latter step is not taken, rapid reinfection occurs.

Examine the slide showing a pinworm under scanning or low power.

8. Ticks

Ticks are arachnids belonging to the same group as mites, spiders, and scorpions. Ticks are ectoparasites on a wide variety of warm-blooded animals, including man, domestic animals, and pets. After mating on a host's body, the female drops to the ground and lays from 3,000 to 6,000 eggs. An egg hatches into larva or a "seed tick." Larvae find a host, feed, drop to the ground, and molt as nymphs. Nymphs feed on a host, drop to the ground, molt, and become adults. The life cycle takes two months to two years, depending on the species. Ticks have infrared sensors on their legs that make the tick attracted to the host's body heat.

Some species of Texas ticks, including *Dermacentor andersoni* (the wood tick), *D. variabilis* (dog tick), and *Amblyomma americanum* (lonestar tick), are vectors of Rocky Mountain Spotted Fever. This serious and eventually potentially deadly disease is caused by a rickettsian, *Rickettsia richettsii*. Rickettsians are smaller than a bacteria and have RNA instead of DNA. They are also intracellular parasites. The rickettsia of an infected female tick invades her eggs; thus all of her offspring are infected. After an average incubation period of seven days, a severe headache, chills, prostration, and muscular pain begins. Fever accompanying the disease may reach 103° to 104°F. On or about the fourth day of the disease, a rash develops on the wrists, ankles, palms, soles of the feet, and forearms. Later, a generalized body rash develops. The disease damages blood vessels, causing brain and heart damage. Starting antibiotic therapy early in the course of the disease has reduced the mortality rate. Fatalities have occurred in the Fort Worth area.

Ticks are also vectors of Lyme disease. The disease was named for Lyme, Connecticut, where it first appeared in 1975. *Borrelia burgdorferi* appears to cause the disease, which is characterized by a skin rash, chills, fatigue, cardiac and neurological disorders, and severe arthritis, Treatment involves appropriate antibiotic therapy. Treatment must be early when the first signs appear—otherwise the disease will take its toll. The disease first appeared in Tarrant County in the summer of 1984.

Examine the slide of a tick under the dissecting microscope.

9. Complete Table 23.B.1.

TABLE 23.B.1   Results

Common Name	Scientific Name	Name of the Disease	How is the Parasite Acquired	Intermediate Host	Definite Host	Methods of Prevention
1						
2						
3						
4						
5						
6						
7						
8						

23 LABORATORY

Survey of the Kingdoms

A. Instructional Objectives:

1. Illustrate understanding of taxonomic principles by explaining:
   a. Why organisms placed in the same taxonomic group are thought to be related.
   b. The difference between taxonomy and binomial nomenclature.
   c. The relationship between an organism's scientific name and binomial nomenclature.
   d. The concept of a species.

2. Illustrate knowledge of each kingdom by:
   a. listing the major characteristics of each Kingdom.
   b. Being able to list and give the characteristics of each Kingdom's taxonomic subdivisions.
   c. Being able to visually recognize the organisms listed at the end of the exercise, and being able to place them in the appropriate taxa.

3. Define these terms: prokaryotic cell, eukaryotic cell, unicellular, multicellular, protozoa, alga, hyphae, heterotroph, saprotroph, decomposer, aquatic, terrestrial, vascular plant, xylem, phloem, invertebrate, vertebrate.

B. Introduction

Taxonomy is defined as an orderly classification or arrangement of various plants and animals. The plant and animal kingdoms are divided into the following subdivisions in descending order: Kingdoms, Phyla, Classes, Orders, Families, Genera (singular is Genus), Species. Organisms placed in the same taxon (plural is taxa) are related by a common evolutionary history and shared genes. The last two classification names, Genus and Species, constitute the organism's name. The Genus name is a Latin noun, while the Species name is a Latin adjective. When hand written or typed, a scientific name is underlined. On the printed page, the name is placed in italics. Man's scientific name is *Homo sapiens*

L. The letter "L" stands for Linnaeus, the man who is considered the founder of taxonomy. He developed the system of binomial nomenclature, or the assigning of a Genus and Species name to all organisms. A Species consists of a group of freely interbreeding individuals that produce viable, fertile offspring. Species sharing characteristics in common are placed in the same Genus. Related Genera constitute a Family. Related Families comprise an Order, related Orders make up a Class, and related Classes are a Phylum. Related Phyla are placed in the same Kingdom, as they share certain characteristics in common. For example, compare the taxonomy of man and the house cat:

Man	House cat
Kingdom: Animalia	Animalia
Phylum: Chordata	Chordata
Class: Mammalia	Mammalia
Order: Primata	Carnivora
Family: Homonidae	Felidae
Genus: *Homo*	*Felis*
Species: *sapiens*	*domestica*

Man and the house cat are placed in the same Kingdom, Phylum, and Class, because both share certain characteristics. For example, an animal is a multicellular organism with membranous cell walls, requires oxygen and organic nutrients, and is capable of voluntary movement. The Phylum name is chordate because both have a dorsal hollow nerve tube, a notochord (embryonic forerunner to backbone), and pharyngeal gill slits (in embryonic development). They are also placed in the same Class because as mammals, they have hair and produce milk from mammary glands. They do not share the same Order because they do not share crucial or diagnostic characteristics in common at that level, or below it. The house cat is a carnivore because it typically eats only meat, or meat products. Man is placed in the primate Order because it represents the most highly developed Order of mammals. It is clear from looking at the example above that a substantial amount of information can be gathered by studying the taxonomic scheme. Cat and man are related, and they share genes in common, but the cat is not the most closely related animal to man. Most authorities agree on the classic five-Kingdom system, developed by Whittaker. They are the Kingdoms Monera, Protista, Fungi (now increasingly called Mycota), Plantae, and Animalia. A higher taxonomic classification above the Kingdom is Domain, of which there are three—Archaea, Bacteria, and Eukarya.

C. Materials

Look for the various specimens in the taxonomy set. They will be set out in the lab.

D. Methods

Carefully read the information included in this lab. Try to understand and remember what characteristics place a particular organism in the correct category. Complete the chart found in Part E of the lab.

Kingdom: *Monera*

Organisms placed in this Kingdom are prokaryotic cells. These cells do not have a membrane-bound nucleus, and have few membranous organelles. Compared to the other Kingdoms, these organisms have simple internal structures. Observe Figure 23.1. They are usually unicellular, or colonial. If colonial, it means they associate by clustering or grouping together to form a colony. They include bacteria and blue-green algae (aka cyanobacteria). Figure 23.2 is an example of cyanobacteria, aka blue-green algae. Bacteria are found everywhere. There

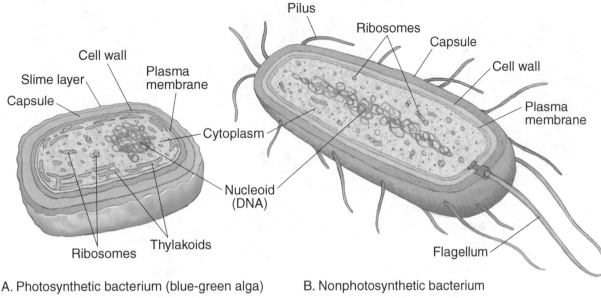

A. Photosynthetic bacterium (blue-green alga)   B. Nonphotosynthetic bacterium

FIGURE 23.1   Kingdom Monera

Copyright © Kendall/Hunt Publishing Company

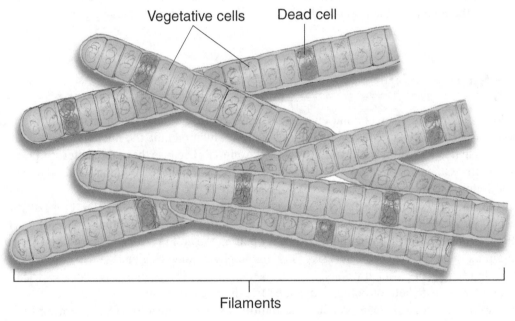

FIGURE 23.2   Cyanobacteria

Copyright © Kendall/Hunt Publishing Company

are billions in our mouth, on our skin, in our colon, in the soil, and so on. Many bacteria are called germs because they cause diseases such as boils, lock-jaw, syphilis, strep throat, and tooth decay. Some produce antibiotics, such as erythromycin and streptomycin. Many remove gases from the atmosphere. They get their energy from a process called chemosynthesis, photosynthesis, or from the decay process.

True bacteria (eubacteria) are members of the Domain Eubacteria, or just known as Bacteria. They reproduce by asexual reproduction, known as binary fission. Some exchange DNA by a process called conjugation. Their cells have three distinct shapes: bacillus (rod shaped), cocci (round, spherical shaped), or spirilla (spiral, corkscrew, helical shaped). Study Figure 23.3.

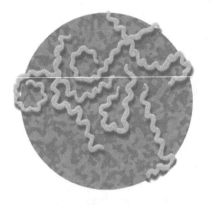

Spirillum
(corkscrew-shaped)

Bacillus
(rod-shaped)

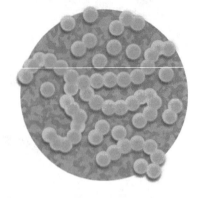

Coccus
(spherical)

FIGURE 23.3   Bacteria

Blue-green algae used to be considered an algae, which is a type of plant. Today they are classified as members of the Phylum Cyanobacteria. Examples (from slides) are *Nostoc* and *Oscillatoria*.

A new Phylum is called Archaea. The bacteria classified in this Phylum differ in cell wall structure, membrane lipid structure, and protein synthesis. They are found living in extreme habitats, such as boiling hot springs, the bottom of the ocean in deep-sea thermal vents, inside volcanoes, or in extremely salty waters.

Kingdom: *Protista*

Kingdom Protista consists of unicellular eukaryotes. Eukaryotic cells are much larger and more complex than prokaryotes. They have membrane-bound organelles, which include the nucleus. Some of the unicellular organisms in this Kingdom were once thought to be animals because they were heterotrophic, and were called protozoans. Others, containing chloroplasts, were once thought to be plants, and were called algae. They are both uni- and multicellular. Study Figure 23.4.

The protozoa phyla were classified on the basis of locomotor devices, flagella, cilia, or pseudopods (false feet). Flagella are long, single extensions from the cellular plasma membrane, made of a cytoskeletal protein. Flagella move in a whip-like motion, thus propelling the cell in a forward movement. Examples of phyla containing these types of protistans are: (1) Euglenophyta (slide Euglena)—also contains an eyespot, and are found in fresh, but polluted waters. (2) Zooflagellates (slide Trypanosoma, causes African sleeping sickness), all of which are heterotrophic, and can be free-living or parasitic.

Cilia, also made of cytoskeletal proteins, are multiple, short extensions of the plasma membrane that beat in a back-and-forth, wavelike motion. The beating sets up a current in the fluid that surrounds the cell, and the cell turns and moves as a result. One example of a phylum containing cilia for movement is: (1) Ciliophora (slide Paramecium). The paramecium is slipper-shaped, covered on the outside with cilia, and may contain two different-shaped nuclei, a macronucleus and a micronucleus.

Other phyla use a "false-foot," or pseudopod, for motion. For example, Phylum Sarcodina (slide Amoeba) is a classic example of an organism that moves by extensions of the pseudopod. Another example is Phylum Sporozoan (slide Plasmodium). The plasmodium is a blood parasite that causes malaria in man, from a female mosquito bite. Sporozoans are parasites with an asexual stage that produces spores.

To start with the protistan algae you will study one group, collectively called the diatoms (1), that belongs to Phylum Chrysophyta. Members of this phylum are golden brown algae. Their cell walls contain silica, and show wide variation in shape and sculpturing. Observe Figure 23.5. Diatoms are the most common

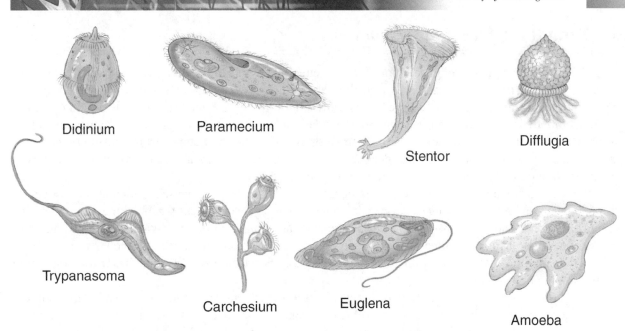

FIGURE 23.4 Kingdom Protista
Copyright © Kendall/Hunt Publishing Company

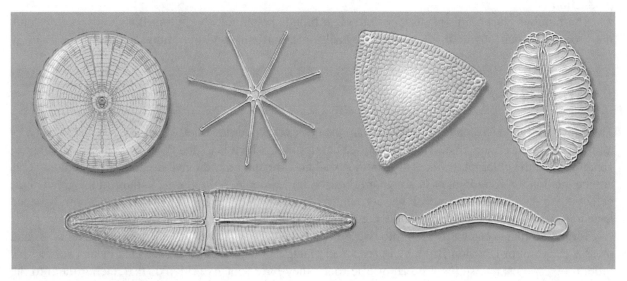

FIGURE 23.5 Diatoms
Copyright © Kendall/Hunt Publishing Company

marine organisms, and are at the bottom of the marine food chain, otherwise known as plankton. Their remains form deposits of diatomaceous earth.

Other protistan algae are (2) Phylum Chlorophyta, known as "green algae" because they contain chlorophyll. Slide examples are Spirogyra and Volvox. The third group, the dinoflagellates (3), are also known as Phylum Pyrrophyta, or "fire plants." Their cell wall is made of cellulose plates, and if disturbed, they luminesce. They are responsible for "red tides," which are destructive to habitats because they kill massive amounts of fish. They give off toxic waste products in the water during rapid growth periods, called "blooms." An example is the slide Ceratium. The last two examples are seaweed, Phylum Rhodophyta (4), known as "red algae" because they contain two pigments, one red and one blue, and Phylum Phaeophyta (5), or "brown algae," which are multicellular and marine.

Kingdom: *Fungi or Mycota*

Members of the Kingdom Fungi are spore-forming eukaryotes, which obtain nutrients from decomposing organic matter. The germinated spores (after the absorption of water) produce filaments called hyphae. A group of branching hyphae is referred to as a mycelium. Fungi are heterotrophs, meaning they cannot manufacture their own food, and must obtain it by other means. They secrete digestive enzymes into a food source to break it down, and then absorb the small products. If the food comes from a dead organism, the fungus is a saprotroph, and is termed a decomposer. Other fungi are parasites that feed on living things by modified hyphae called haustra, which grow into the host's tissue.

Some fungi are mutualistic, forming an association with algae called lichens. The mutualistic association is beneficial to both fungi and algae, because it allows lichens to live in places where neither fungi nor algae could live alone. The algae provide food from photosynthesis, and the fungi provide anchorage and retain water. One kind of fungus, called *Penicillium,* produces the antibiotic penicillin. It belongs to Division or Phylum Deuteromycota, called "imperfect fungi" because they lack a sexual stage. Their reproduction is asexual, and produces spores.

Yeast (baker's yeast), another kind of fungus, is used to ferment carbohydrates into ethyl alcohol, and to produce carbon dioxide bubbles to causebread dough to rise. It is classified in Division or Phylum Ascomycota (sac fungi), which also contains morels and truffles. Mycorrhizae is a term that represents a mutualistic association between fungi and plant roots, allowing both to thrive in the environment.

Black bread mold (Phylum Zygomycota) belongs to the Genus *Rhizopus.* The mycelia grow on the surface of bread or fruit. Some of the hyphae showing vertical growth are modified as sporangia for spore production.

The familiar mushroom (genus *Coprinus*) is a fungus belonging to Phylum Basidiomycota or "club mushrooms." Other fungi in this Kingdom are toadstools, puffballs, and shelf fungi (on trees).

Kingdom: *Plantae*

The Kingdom Plantae is made of multicellular eukaryotes that conduct photosynthesis. For an overview of all plants, consult Figure 23.6. All plants utilize chlorophyll, which is a green pigment contained in an organelle called the chloroplast. Plants use both sexual and asexual reproductive modes. Some aquatic algae are included in the Kingdom, or in Kingdom Protista. For example, *Spirogyra* is a filamentous green algae belonging to Phylum Chlorophyta. It gets its name from the fact that the chloroplast displays a spiral-shaped appearance, and can be easily seen under the microscope. An example is shown in Figure 23.7.

Terrestrial plants can be divided into two groups—the bryophytes and tracheophytes. Liverworts and mosses (Phylum or Division Bryophyta) are small plants lacking vascular tissue. This means no, or very little, roots, shoots, and stems. Instead of true roots, they have root-like rhizoids that anchor them to the soil. *Mnium* is a moss. They require water for fertilization, and because they don't have vascular tissue, they are restricted to moist habitats.

Tracheophytes do contain vascular tissue, which will conduct materials within the body of the plant. Vascular tissue called xylem will conduct water and minerals from the soil up through the stem and to the leaves, against the flow of gravity. Phloem is vascular tissue that will conduct hormones and sugar solutions from the leaves (where photosynthesis takes place) down through the stem, and in some plants, to the roots. Primitive tracheophytes do not have seeds, but instead produce spores. An example is the fern *Polyplodium* (subphylum Pterophyta). The fern's gametophytes can be seen in slides, labeled antheridia (male) and archegonia (female). More advanced tracheophytes produce seeds, of which gymnosperms and angiosperms are examples. Gymnosperms means "naked" or uncovered seed, and they are the familiar pine, with the pine cone containing the seed. They are placed in Phylum Coniferophyta. Other examples include spruce, fir, and hemlock trees. They are commonly referred to as "evergreens." Angiosperms means "covered seed," and the familiar flowering plants (Phylum Anthophyta) are examples. The angiosperms are divided into two important classes, called monocots and dicots. The terms

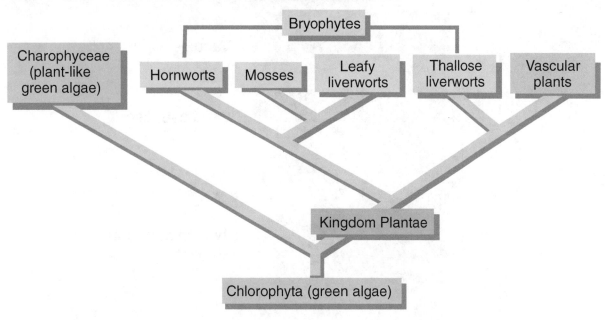

FIGURE 23.6 Overview of Plant Kingdom
Copyright © Kendall/Hunt Publishing Company

come from a shortened version of the word "cotyledon." It is an area in the plant embryo, which serves to store food. It is a leaf-like structure found in the seed, and is also known as the seed leaf—monocots have one; dicots have two. Monocots have a parallel leaf vein pattern; dicots have a branched or netted vein pattern on the leaf. Monocot flower parts are found in multiples of three; dicot flower parts are in multiples of four or five. The stems of monocots are herbaceous, and the stems of dicots are either herbaceous or woody. Examples of monocots are palm, lily, orchid, grasses, sugar cane, corn, wheat, and rice. Examples of dicots are roses, buttercup, lima beans, peas, oak, and maple. Gymnosperms and angiosperms are economically important because we use them for food, fiber, lumber, and medications. Observe Figure 23.8.

Kingdom: *Animalia*

Animals are clearly multicellular eukaryotes, heterotrophic and motile. Animals cannot manufacture their own food, and digestion is internal. Several Phyla constitute the Kingdom Animalia. For an overview of all animals, study Figure 23.9. Animals without a backbone are called invertebrates. Sponges (Phylum Porifera) are aquatic, filter-feeding animals that contain a protein called spongin that makes up the skeleton of sponges. Other invertebrates such as jellyfish and coral (Phylum Cnidaria) are aquatic organisms with tentacles that contain stinging cells called cnidocytes. The cells are used to sting and paralyze their prey. A third phylum is that of the flatworms, called Platyhelminthes. Examples are planaria (slide), flukes, and tapeworms. Phylum Nematoda describes roundworms that are unsegmented. An example is Ascaris. You may have a slide showing the cross-section of a male and female, with the diameter of the female being much larger than the male. Phylum Annelida contains round, segmented worms, the best example of which is the earthworm, Class Oligochaeta. Also included in this Phylum are marine worms, sand worms, Class Polychaeta, and leeches, Class Hirudinea.

Other examples of invertebrates are squid, octopus, and cuttlefish. They have a modified foot in the form of tentacles. They are classified in Phylum Mollusca, Class Cephalopoda. Clams, scallops, oysters, and mussels are bivalves with two shells (bivalves) protecting their soft body and muscular foot. We eat the entire body of an oyster. Gastropods are snails and slugs, whelks, and conch shells. Chitons are classified in Class Polyplacophora. They are found in intertidal areas.

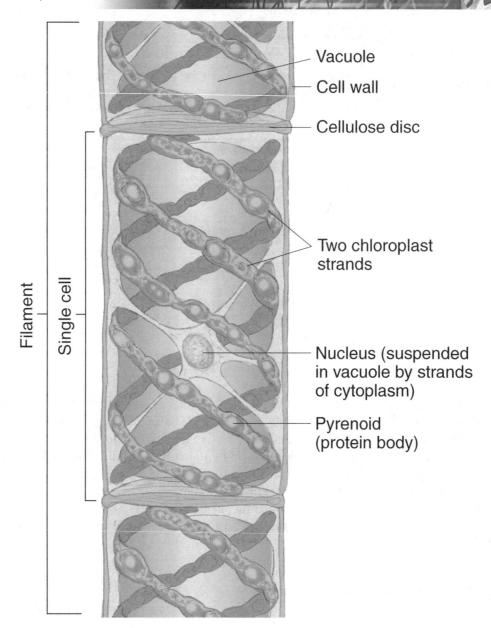

FIGURE 23.7   Spirogyra
Copyright © Kendall/Hunt Publishing Company

Phylum Crustacea, Class Crustacea contain shrimp, crab, lobster, and crayfish.

Members of Phylum Arthrophoda are extremely diverse and, in fact, approximately 75% of all animals are members of this phylum. These animals all have a body structure consisting of a head, thorax, and an abdomen. In some cases, the thorax and head are joined together. They have jointed legs and an exoskeleton composed of a protein called chitin. Centipedes, millipedes, insects, spiders, ticks, scorpions, and the horseshoe crab are all arthropods.

Echinoderms (Phylum Echinodermata) are marine animals with spiny skins, and are considered the most advanced examples of the invertebrates. Starfish are an example, along with sea cucumbers, sea urchins, and brittle stars.

Animals classified as chordates are the most advanced in Kingdom Animalia. For an overview of chordates, consult Figure 23.10. Animals in Phylum Chordata have a notochord (embryonic forerunner to the vertebrae), a dorsal hollow nerve tube (embryonic forerunner to the spinal cord), and pharyngeal gill slits. Vertebrates are chordates that develop a backbone. Jawless fishes are classified in Class Agnatha. An example

Dicot (two cotyledons)

| Pollen grains have three pores or furrows | Seeds have two cotyledons | Flowers have four or five floral parts (or multiples thereof) | Leaves are oval or palmate, with net-like veins | Vascular bundles arranged in a ring around stem | Tap roots |

Monocot (one cotyledon)

| Pollen grains have one pore or furrow | Seeds have one cotyledon | Flowers have three floral parts (or multiples thereof) | Leaves are narrow, with parallel veins | Vascular bundles small, and spread throughout stem | Fibrous roots |

FIGURE 23.8   Flowering Plants

Copyright © Kendall/Hunt Publishing Company

would be the hagfish. Examples of marine animals in this Phylum are sharks and rays, with an endoskeleton composed of cartilage (Class Chondrichthyes). Bony fishes (Class Osteichthyes) are goldfish, clownfish, perch, mackerel, trout, salmon, and so on. Frogs, toads, and salamanders are in Class Amphibia. They have moist skin, and require water for reproduction, as sexual reproduction is external. Turtles, snakes, lizards, crocodiles, and alligators are reptiles (Class Reptilia), and have epidermal scales, dry skin, and internal fertilization. They are cold-blooded, which means their body temperature is dependent upon their environment. Many snakes are poisonous. Birds (Class Aves) have a body temperature independent of the environment (warm-blooded), feathers, and two legs. Mammals (Class Mammalia) are warm-blooded, have hair, use internal fertilization for sexual reproduction, and produce milk from mammary glands for their young. Examples are dogs, cats, cows, sheep, horses, bats, dolphins, whales, and humans.

Examine the microscope slides, and preserved specimens in the lab, and complete the following study tables: They represent bacteria, protists, fungi, plants and animal diversity. Use the textbook and/or lab manual when in doubt. Classify the organism as completely as possible, noting the characteristics of each one.

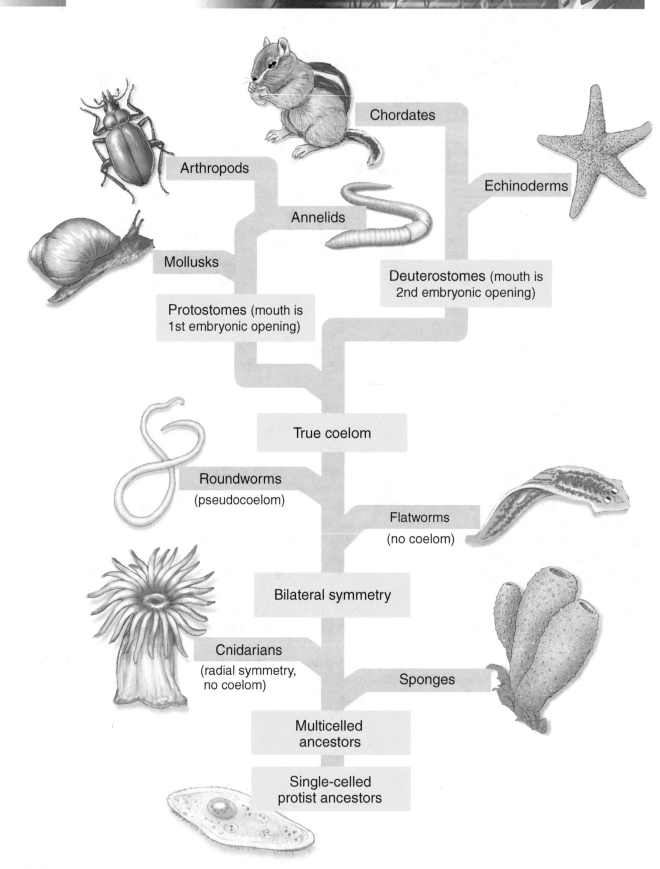

FIGURE 23.9   Animal Family Tree
Copyright © Kendall/Hunt Publishing Company

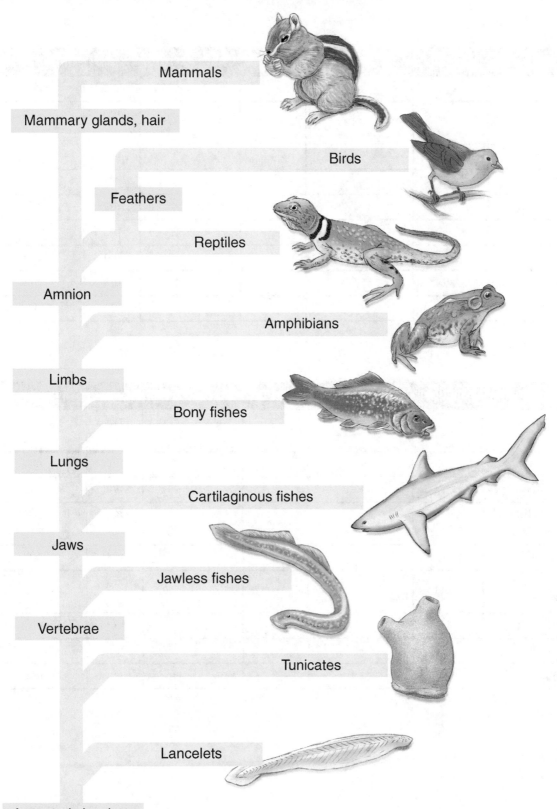

Mammals

Mammary glands, hair

Birds

Feathers

Reptiles

Amnion

Amphibians

Limbs

Bony fishes

Lungs

Cartilaginous fishes

Jaws

Jawless fishes

Vertebrae

Tunicates

Lancelets

Ancestral chordates

FIGURE 24.10 Chordate Family Tree

Copyright © Kendall/Hunt Publishing Company

TABLE 23.1

Organism	Kingdom	Characteristics	Phylum	Characteristics	Class	Characteristics
E. coli					n/a	n/a
Helicobacter pylori		"			n/a	n/a
Nostoc		"			n/a	n/a
Amoeba					n/a	n/a
Spirogyra		"			n/a	n/a
Red algae		"			n/a	n/a
Fucus		"			n/a	n/a
Plasmodium		"			n/a	n/a
Euglena		"			n/a	n/a

TABLE 23.2

Organism	Kingdom	Characteristics	Phylum	Characteristics	Class	Characteristics
Trypanosoma bruceii		"			n/a	n/a
Paramecium		"			n/a	n/a
Diatoms		"			n/a	n/a
species that cause red tides		"			n/a	n/a
Rhizopus					n/a	n/a
Sacchromycete cereversii		"			n/a	n/a
Mushroom		"			n/a	n/a
Penicillium		"			n/a	n/a
Mnium					n/a	n/a

TABLE 23.3

Organism	Kingdom	Characteristics	Phylum	Characteristics	Class	Characteristics
Lycopodium		"			n/a	n/a
Equisetum		"			n/a	n/a
Fern		"			n/a	n/a
Cycad		"			n/a	n/a
Pine tree		"			n/a	n/a
n/a					Staminate cone	
					Ovate cone	
					Mature cone	
Lily		"				
Bean		"	"	"		

TABLE 23.5

Organism	Kingdom	Characteristics	Phylum	Characteristics	Class	Characteristics
Leech		"		"		
snail		"				
squid		"		"		
chiton		"		"		
clam		"		"		
centipede		"				
crayfish		"		"		
dragonfly		"		"		
tick		"		"		

TABLE 23.6

Organism	Kingdom	Characteristics	Phylum	Characteristics	Class	Characteristics
starfish		"			n/a	n/a
lamprey eel		"				
dogfish shark		"		"		
perch		"		"		
salamander		"		"		
snake		"		"		
sparrow		"		"		
cow		"		"		

24

LABORATORY

Flowers and Fruits

☐ Part A. Flowers

Instructional Objectives:
1. Be able to identify the following floral structures and their function from a model of the flower or from live specimens:
   a. peduncle
   b. receptacle
   c. sepals (calyx)
   d. petals (corolla)
   e. carpel (pistil)
      i. ovary
      ii. style
      iii. stigma
   f. stamen
      i. anther
      ii. filament
2. Be able to distinguish between the following:
   a. perfect and imperfect flower
   b. complete and incomplete flower
   c. monocot and dicot

Materials

Flower model

Live specimens of flowers

Dissecting microscope

Introduction

Flowering plants, the angiosperms, are the dominant life form of Kingdom Plantae on Earth. Today, angiosperms make up 80% of the species of all living plants. The reason for their dominance is their ability to rapidly reproduce and their independence from water-borne reproduction. Mosses and ferns are dependent on living in a moist environment for their reproductive success. Though gymnosperms are not dependent on water for reproduction, it takes two growing seasons for the formation of the seed. Also, gymnosperms depend on wind aided-pollination.

Flowering plants exhibit a tremendous amount of diversity. Grasses, herbs, vines, shrubs, and trees are all types of flowering plants. The majority of foods (cereal grains, vegetables, and fruits) that we eat are the products of flowering plants.

The flower is more than a showy extravagance of nature, for it is the means of sexual reproduction for the plant. Over millions of years insects and flowering plants have coevolved. Some flowering plants are totally dependent on pollinators (bees, wasps, hummingbirds, and bats) for their success in reproduction. The domestic honeybee population is in decline. The decline of the honeybee is called colony collapse disorder. The reason for the collapse is not known as of the present (2007), but the indiscriminate use of pesticide may have weakened the animal to become susceptible to disease. The result is that many of the plant products that are totally dependent on bee pollination that we eat could become too expensive. For example, Asian pears could become a delicacy since honeybees in China are disappearing.

Plants that have showy flowers and a scent are more likely to attract pollinators. Plants that have small, inconspicuous flowers with no scent are more likely to be wind-aided in pollination. Hummingbirds are attracted red tubular flowers with little or no odor, while butterflies are attracted to tubular flowers that have a strong, sweet odor. Bees are attracted to symmetrical flowers that form a rosette of petals to form a landing platform, and these flowers have white, yellow, orange, or blue petals. Also these flowers have a sweet, fragrant scent.

Angiosperms can be classified as either monocots or dicots. Monocots have floral parts in multiples of threes, while dicot floral parts are in multiples of fours or fives. Monocots also have parallel venation whereas dicots have a netted venation (pinnate or palmate).

Flowers are classified as being complete or incomplete. A complete flower has all floral parts: sepals, petals, stamens, and carpels. An incomplete flower is missing one or more of these floral parts. A perfect flower has both male (stamens) and female (carpels) reproductive structures. An imperfect flower has either stamens or carpels.

Procedure

1. Examine fresh flowers of five different species, preferably with different floral characteristics.

2. Identify the parts of each flower.

3. Sketch in the place provided any flower shapes or structures needed.

4. Complete Data Table 24.A.1, Flowers.

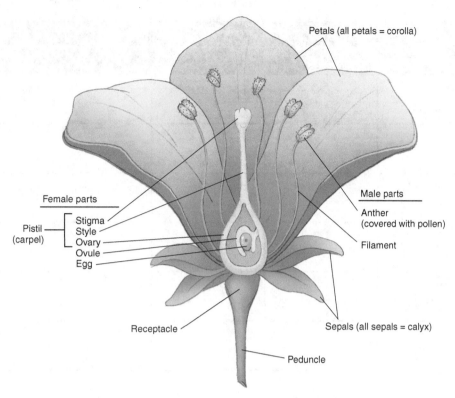

FIGURE 24.A.1 Generalized Structure of a Dicot Flower
Copyright © Kendall/Hunt Publishing Company.

TABLE 24.A.1 Floral Parts

Part	Description/Function
Pedicel	Stalk that supports the flower
Receptacle	Center of the base of the flower; end of the stem; flower parts are attached here
Sepal	Small and leaf-like; different in shape from the foliage leaves; may be green, brown, or colored like the petals; major function is protection of the immature inner parts
Calyx	All of the sepals together
Petal	Brightly colored or white and very broad; serves as additional protection in the early stages; attracts pollinators; serves as landing pad for pollinators; some have glands (nectaries) at the base to produce sweet nectar or volatile oils for scent
Corolla	All the petals together
Stamen	Pollen-bearing structure, composed of filament and anther
Filament	Thin stalk, supports the anther
Anther	Produces pollen
Carpel	Female reproductive structure; composed of stigma, style, and ovary; often pear-shaped; located at the center of the flower
Stigma	At the tip of the carpel; produces a sticky substance that captures the pollen grains; important in the germination of pollen
Style	Connects the stigma to the ovary; often long and narrow; pollen tubes grow toward the ovules
Ovary	Base of the carpel; protects ovules; matures to form the fruit

DATA TABLE 24.A.1   Flowers

Features of the flower	Name of Plant			
	1	2	3	4
Number of petals				
Number of sepals				
Floral parts missing				
Complete/incomplete				
Perfect/imperfect				
Color				

Name: _____

QUESTIONS

1. The flower below had the following floral parts present:

Floral Parts	Present (+)/ Absent (−)
Sepals	+
Petals	+
Pistils	+
Stamens	−

The flower is classified as (complete/ incomplete) _____ and (perfect/imperfect) _____.

2. The flower below had the following floral parts present:

Floral Parts	Present (+)/ Absent (−)
Sepals	−
Petals	−
Pistils	+
Stamens	+

The flower is classified as (complete/ incomplete) _____ and (perfect/imperfect) _____.

3. The number of floral parts of the flower below were counted:

Floral Parts	Number
Pistils	9
Stamens	9

The flower is classified as a (dicot/ monocot) _____.

4. If you were in a tropical forest and you came upon a vine with red tubular flowers, what type of pollinator would you expect to be visiting the flowers for nectar?

5. If in a garden you noticed a flower that had bright yellow petals arranged in a rosette, what type of pollinator would you expect to be visiting the flowers for nectar?

6. The success of flowering plants is due to the process of coevolution with which group of animals?

7. The success of flowering plants is the result of which two factors, aside from pollinators?

Part B. Fruits

Instructional Objectives:
1. Be able to distinguish between the following:
    a. a simple and a compound fruit.
    b. a dry and fleshy fruit.
    c. a grain and a nut.
    d. a legume and a capsule.
    e. a drupe, berry, hesparidium, pepo, and pome.

Materials

cherries, apples or pears, peaches, unshelled almonds and pecans, oranges, strawberries, squash, dried beans, green beans, dried corn, okra or peppers, tomatoes, raspberries or blackberries

Introduction

Fruits are formed from the ovary and/or modified receptacle parts. Fruits surround and protect the seeds. Both seeds and the fruit may be involved in the dispersal of the sporophyte embryo. In parthenocarpy, fruits are formed without fertilization. An example of parthenocarpy is seedless grapes.

Fruits are formed from the pericarp. The pericarp is the part of the ovary wall that immediately surrounds the ovule(s). The three layers of the pericarp are:

Exocarp: the outermost layer, often consisting of only the epidermis
Mesocarp: middle layer, which varies in thickness
Endocarp: shows considerable variation between species

Fruits are classified as either simple or compound. A simple fruit forms from a single ovary consisting of one or two carpels. Compound fruits may be either multiple or aggregate. A multiple fruit forms from the ovaries of many flowers—for example, mulberry, sycamore, pineapple. An aggregate fruit forms from several ovaries in one flower—for example, strawberry, magnolia, raspberry.

Simple fruits may be dry or fleshy. At maturity, dry fruits may have only one seed or have many seeds. Fruits with one seed are grains or nuts. Grains are fruits of cereals. Cereals include rice, wheat, corn, barley, and oats. Grains have a dry pericarp fused with the seedcoat. Most of the seed is the endosperm, which is composed of plant starch. The outer layer of the endosperm cells forms the aleurone layer. This layer contains most of the seed's protein and fat. Bran is the aleurone layer and the embryo plant. When you eat white bread and white rice, you are robbed of this important part of the seed. Whole-wheat bread and brown rice are more nutritious than white bread and white rice. Bran helps reduce serum cholesterol.

Nuts are fruits that have a hard or woody pericarp. The fleshy edible seed can be separated from the pericarp. Examples are pecans, acorns, walnuts, and hickory nuts. The fleshy seed of nuts contains a great amount of oil.

Simple fruits that vary from two to many seeds may be classified as capsules or legumes. Capsules are fruits that have an ovary with several cavities (when sliced in a cross-section) and have several to many seeds. Okra and the jalapeño pepper are examples of capsules. Legumes contain a mature ovary with a single cavity. Commonly this is called a pod. Legumes are peanuts, beans, peas, lentils, soybeans, and chickpeas. Black-eye peas are actually beans. A green bean or sugar snap pea is eaten before the fleshy pericarp has dried out. When dry, the pod splits down both sides to release the seeds.

Fleshy fruits are drupes, pomes, berries, and pepos. Drupes, berries, and pomes have a fleshy pericarp. Drupes are simple fleshy fruits that are composed of an ovary with one seed. The exocarp is the skin of the fruit. The mesocarp is the flesh of the fruit. The seed is surrounded by a very hard stone—the endocarp. Drupes include cherries, almonds, peaches, plums, and apricots. The exocarp and the mesocarp have been removed before the almond reaches the store. Also the stony endocarp must be removed to eat the seed, the almond.

Berries form from an ovary containing many seeds that are embedded in a fleshy endocarp and mesocarp. The skin of a berry is the exocarp. Tomatoes are true berries. In 1893, the United States Supreme Court made a decision that tomatoes are a vegetable. The reason was to make tomatoes subject to an import duty tax.

Citrus fruits are a type of berry called a hesparidium. Citrus fruits include lemons, limes, oranges, and grapefruit. The leathery peel is the exocarp and the mesocarp. The pulpy segments are the endocarp. The juice sacs are cellular outgrowths from the endocarp. Seeds are found between the juice sacs. The membrane between the sections of an orange is the carpal wall.

The pepo is the fruit of cucumbers, watermelons, squashes, and pumpkins. The rind is the fused receptacle tissue with the exocarp. The flesh is the mesocarp and endocarp.

A pome is a fleshy fruit that develops in part from the surrounding tissue of the flower (base of sepals and petals). The ovary wall is seen as the "core" around the seeds. A pome is the fruit of an apple, pear, or rose. Rose hips are pomes. Most of the eaten flesh is the receptacle tissue. The core is the ovary.

Procedure

1. Examine fruits and seeds on display.
2. Classify them according to the following scheme:
   a. Is the fruit simple or multiple?
   b. If the fruit is multiple, is it compound or aggregate?
   c. If the fruit is simple, is it dry or fleshy?
   d. If the fruit is dry, does it have only one seed or many seeds?
   e. If the fruit has only one seed, is it a grain or a nut?
   f. If the fruit has many seeds, is it a capsule or a legume?
   g. If the fleshy fruit is a drupe, does the ovary have one seed that is surrounded by a very hard stone?
   h. If the fleshy fruit is a berry, is the ovary soft and fleshy and does the ovary contain many seeds?
   i. If the fleshy fruit has a leathery exocarp and mesocarp, is it a hesparidium?
   j. If the fleshy fruit is a pepo, does it have a rind made of exocarp and the flesh made of endocarp and mesocarp?
   k. If the fleshy fruit is a pome, does it have a core with seeds?

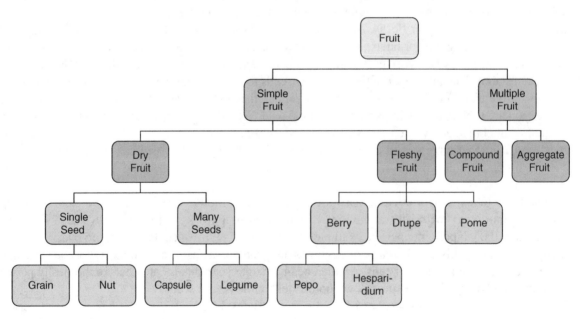

FIGURE 24.B.1   Classification of Fruits
Copyright © Kendall/Hunt Publishing Company.

3. Complete Data Table 24.B.1, Classification of Fruits.

DATA TABLE 24.B.1 Classification of Fruits.

Place an X in the square for each characteristic that applies to the following fruits:

Item	Pecan	Almond	Dry Bean	Green Bean	Tomato	Dry Corn	Almond	Orange	Squash	Jalapeño or Orka	Apple	Peach	Strawberry
Fruit													
Seed													
Simple													
Compound													
Dry													
Fleshy													
Pod													
Grain													
Capsule													
Nut													
Drupe													
Berry													
Pome													
Pepo													
Hespaeridium													

Name: _____

QUESTIONS

1. How may fleshy fruits be dispersed?

2. How may dry fruits be dispersed?

3. Describe the difference between a drupe and a berry.

4. Describe the difference between a grain and a nut.

5. Describe the difference between a legume and a capsule.

6. Describe the difference between a hesparidium and a pepo.

7. Describe the difference between an aggregate and a compound fruit.

25
L A B O R A T O R Y

South Campus Plants

Instructional Objectives:
1. Be able to distinguish between the following:
   a. shrub and tree.
   b. herbaceous and woody plant.
   c. parallel, pinnate, and palmate venation.
   d. evergreen and deciduous plants.
   e. simple leaf and compound leaf.
   f. alternate leaf attachment and opposite leaf attachment.
   g. staminate cone and pistillate cone.
   h. entire (smooth) margin and serrated (toothed) margin.
   i. gymnosperm and angiosperm.
2. Be able to identify the following plant structures:
   a. blade.
   b. petiole.
   c. rachis.
   d. leaflet.
   e. axillary bud.
3. Be able to identify the plants in Table 25.1 and recognize their distinguishing characteristics.

☐ Introduction

This exercise is designed to acquaint you with some selected plants on the South Campus of Tarrant County College. Some of these plants are woody plants, either trees or shrubs, while others are herbaceous plants. Some of these plants are evergreen while others are deciduous. Some of these plants produce flowers while others form cones. The plants have been selected because they are found around Fort Worth area homes.

First we need to distinguish between woody and herbaceous plants. **Woody plants** have a trunk and stems covered with bark. Trees and shrubs are woody plants. **Trees** are large woody plants having a large single woody stem, called a trunk. The first lateral

TABLE 25.1   South Campus Plants

Scientific Name	Common Name	Location
Quercus shumardii	Red Oak	Southeast entrance of Science Building
Ulmus crassifolia	Cedar Elm	
Ilex vomittoria	Yaupon Holly	East side of Science Building
Setcreasea purpurea	Purple Heart	East side of Science Building
Nandina domestica	Nadina or Heavenly Bamboo	Northeast entrance of Science Building
Fraxinus pennsylvannia	Green Ash	North side of Science Building
Lagerstroemis indica	Crepe Myrtle	West and south side of the Rotunda
Pyrus callerya	Bradford Pear	Northeast entrance of Science building
Pinus	Pine	Southeast of Library
Photinia fraseri	Fraser's Red Tipped Photinia	East side of Auto Building
Magnolia grandiflora	Southern Magnolia	East of Library and south of the Rotunda
Taxodium distichium	Bald Cyprus	North of Science Building
Quericus virginia	Live Oak	South of Health Sciences
Hesperaloe parviflora	Red Yucca	Southwest side of Auto Building
Ginkgo biloba	Maidenhair Tree	
Chilopsi linearis	Desert Willow	West side of Health Science Building

branches usually begin some distance from the ground. Shrubs may have multiple trunks. The branches of shrubs are numerous and near the soil's surface. Shrubs may be trained to have only one trunk. These shrubs are referred to as being tree-form. Herbaceous plants are nonwoody, stemmed plants.

In this laboratory exercise we will use type of seed formation, leaf shape, leaf arrangement, and floral arrangement to help identify the plants in this exercise. Since flowers are short-lived, they may not be present for identification.

Seed plants may be classified as either gymnosperms or angiosperms. Gymnosperms are cycads, ginkgos, gnetophytes, and conifers. The name comes from the Greek words *gymno* meaning "naked" and *sperma* meaning "seed." The name gymnosperm means "naked seed." The two types of gymnosperms on campus are the ginkgo and conifers. Both are not closely related to each other.

Angiosperms are the dominant plant life on this earth. Angiosperms are flowering plants that have both flowers and fruits. In Laboratory 24, the type of flowers and fruits will be discussed in more depth.

Leaves are metabolically equipped for photosynthesis, the plant's mechanism for using the raw materials water and carbon dioxide in the presence of sunlight and chlorophyll to form sugar (glucose) and oxygen. Leaves vary is size, shape, and form. Leaves may appear as blades, needles, feathers, cups, or tubes. Leaves appear in different colors (due to the types of pigments in the leaf), may have different odors, may be edible, or may produce toxins to prevent being eaten. Some plants shed their leaves in the fall and are termed deciduous. Other plants never shed their leaves at one time and are considered to be evergreen.

Broad-leaf plants have leaves with a flattened blade and a distinct petiole (leaf stalk) attached to the stem or twig (see Figure 25.1 a). The leaf is thin and flattened to maximize the surface area for capturing the maximum amount of sunlight. The leaves are arranged to be perpendicular to the sunlight.

The pattern of veins in a leaf is called venation. Herbaceous plants have parallel veins. These veins run in the same direction toward the tip of the blade (see Figure 25.1 g). Woody plants have a netted venation,

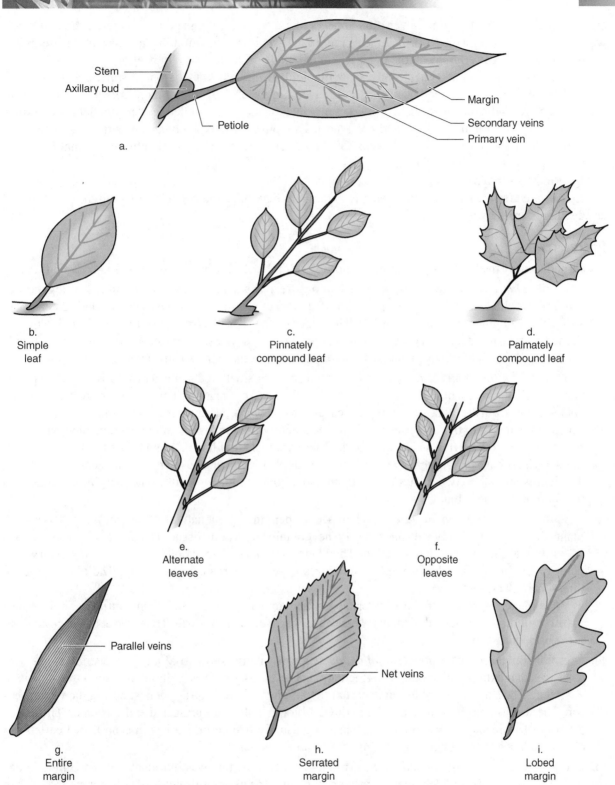

Stem

Axillary bud

Petiole

Margin

Secondary veins

Primary vein

a.

b.
Simple
leaf

c.
Pinnately
compound leaf

d.
Palmately
compound leaf

e.
Alternate
leaves

f.
Opposite
leaves

Parallel veins

Net veins

g.
Entire
margin

h.
Serrated
margin

i.
Lobed
margin

FIGURE 25.1   Leaf Types and Structures

Copyright © Kendall/Hunt Publishing Company.

either pinnate or palmate. Pinnate venation appears like a feather. In pinnate leaves there is a central vein with smaller veins originating off of it (see Figure 25.1). Palmate venation has a finger-like pattern. The veins in a palmate leaf originate at the point where the blade originates with the petiole.

Leaves are classified as either simple or compound. If the leaf blade is one distinct part—that is, is undivided—the leaf is classified as a simple leaf (see Figure 25.1 b). Many monocots have a leaf in the shape of a blade. When the leaf blade is divided into leaf-like sections, called leaflets, the leaf is classified as compound (see Figures 25.1 c and d). True leaves have axillary buds at the petiole-stem junction (see Figure 25.1 a), whereas the leaflets do not. Compound leaves are classified by their venation, either pinnate or palmate (see Figure 21.1).

Leaf margins maybe classified as entire, toothed, or lobed.

Leaves are arranged at the node in simple patterns to maximize the capturing of sunlight. The types of arrangements are shown in Figure 25.1.

Procedure

Locate and identify the following plants and complete Data Table 25.2, Characteristics of South Campus Plants.

1. Red oak, *Quercus shumardii*, is a large deciduous tree. The dead leaves are shed in the spring, not the fall. Prior to the leafing out in the spring, the dead leaves are shed. The leaves have a dark green upper surface and a lighter green under surface. The under surface has a velvety appearance due to downy epidermal hairs that reduce water loss. The leaves have five to seven slender lobes that terminate as bristly tipped teeth. The leaf arrangement is alternate. The fruit of these trees is the acorn.

2. Yaupon holly, *llex vomitoria*, is a large evergreen shrub or small tree that is native to Texas. The species name, while not pleasant sounding, is descriptive, for the American Indians used to make a ceremonial "black brew" from the plant's caffeine-rich leaves. The brew is a purgative that causes the person to vomit. The plant is native to the immediate area. It is frequently used in lawns and may be planted with multiple trunk groupings. Trunks have gray bark. There are male and female plants. The female plants during the winter have red berries. The leaves have a simple margin and have an alternate leaf arrangement. The leaves are one to two inches in length and are about 1/2 inch in width. The margin is toothed and the tip of the blade is blunt.

3. Purple heart, *Setereasea purpurea*, is a herbaceous perennial plant native of Mexico. When herbaceous plants die, they completely wilt since they do not contain any woody tissue. The blades of this plant have an entire margin with parallel venation. The blades are lance-shaped and are pubescent, or hairy. The color of the blades is reddish-purple. The leaves have an alternate arrangement. The plants produce three-petal lilac-like flowers.

   The plants are related to the Wandering Jew and are easily propagated from cuttings. They do best in full sunlight when planted outdoors. This causes more compact growth. They freeze to the ground but grow back each spring.

4. Nandina, or Heavenly Bamboo, *Nandina domestica*, is a native shrub to China. This evergreen has winter leaves that exhibit red to orange pigmentation. The stems, or canes, exhibit erect growth from six to eight feet. The shrubs do best in sun to partial shade. The fruit, red berries, persist during the winter season. The seed is poisonous if eaten. White flowers bloom in the late spring and early summer. The leaves are bi- or tri-pinnate compound leaves. There are usually five entire leaflets in each leaf. Leaflets are elliptic to lance shape. Leaves are arranged in an alternate pattern.

5. The crepe myrtle, *Larerstroemia indica*, is a native of China that was introduced to the United States in 1747. This shrub is widely planted as an ornamental for its beautiful clusters of flowers that bloom for 60 to 120 days in the summer. Flower color may be white or range in many shades of pink, purple, lavender, and red. The fruits that follow are brown or black. The bark is exfoliated from older trunks in thin layers. The underlying bark is cinnamon to gray in color. In early spring the leaves are light reddish-brown. The leaves are simple and have an entire margin. The leaf arrangement is alternate.

6. The cedar elm, *Ulmus crassiflora*, is a deciduous tree native to Texas river bottoms and the limestone hills of the Hill Country. Young trees have a strong vertical growth. The tree has a straight trunk. Mature

trees have a narrow rounded crown with dropping branches. The may reach a height of 80 feet and have a trunk diameter of 2 to 3 feet. The trunk bark is flat, scaly-ridged, and light brown. Trees are drought resistant when firmly established. The leaves are simple and have an alternate arrangement. The leaves are attached to corky stems by short, thick petioles. The margin of the leaves is serrated. The leaves have a dark green upper surface and a pubescent lower surface. Leaves turn yellow before dropping. Flowers are a small, short-stalked clusters in the early fall. The fruit is a small winged seed called a samara.

7. The **Live Oak**, *Quericus virginiana*, is evergreen in warm areas but sheds some of its leaves in the spring. The tree grows to be 40 to 80 feet tall and 60 to 100 feet wide. Some trees live to be 200 to 300 years old. The fruit is an acorn one inch long and 1/2 inch wide. They are produced in clusters from one to five. Animals use them as a food source. The leaves are thick, leathery, and two to five inches in length. The upper surface of the leaf is shiny dark green. The lower leaf surface is a downy, lighter-green color. The margin is entire and the leaf tip is rounded. The leaves are arranged in an alternate pattern.

8. **Fraser's red-tipped photinia**, *Photinia fraseri*, is an evergreen shrub that can make an attractive hedge. In the spring the shrub produces reddish young leaves. The shrub can attain a height of 10 to 15 feet and a width of 15 feet. It may be trimmed to a tree form. The mature simple leaves have a reddish-brown petiole supporting an ovate blade two to three inches in length. The blade terminates into a pointed tip. The margin is sharply serrated. Leaf attachment is alternate. The flowers form in the spring and are dense clusters of small white simple petals. If berries do form, birds and other wildlife do like them.

9. The **Bradford pear**, *Pyrus calleryana*, is an ornamental tree valued for its early spring white flowers and its showy display of coppery-red foliage in the fall. The Bradford pear grows 30 to 50 feet tall and 20 to 30 feet wide. It has a short to moderate life span (25 to 30 years). The leaves are simple, broad leaves with a size of two to five inches. The leaves are round to egg (ovoid) shape. They are alternately arranged. The venation is pinnate. The fruit is small and inedible.

10. The **southern magnolia**, *Magnolia grandiflora*, is an evergreen tree that may reach a height of 60 to 80 feet in the South. The pyramidal crown is dense. Petioles are short and have rusty hairs. The simple leathery leaves are five to eight inches in length. The shape of the leaf is elliptical or ovoid. The margin is entire. The upper surface of the leaf is a dark green and the under surface is rusty to a silvery color. Leaves have a pinnate venation. The leaf arrangement is alternate. The species name comes from the giant white flowers with a six- to eight-inch width. Flowers appear singly from May to June. The flowers produce a pleasing fragrance. The fruit is a follicle containing bright red seeds. This matures in October to November.

11. Pines are coniferous evergreens belonging to Genus *Pinus*. The needle-like leaves occur in bundles called fascicles. Needles are an adaptation to prevent water loss (transpiration). The tree produces small staminate cones, pollen cones. The larger cones are the pistilate cones, female cones. All pollination is wind aided. It takes two growing seasons for a seed to form. The seeds are borne of the scales of the female cone. The seeds have a wing that allows the wind currents to carry the seed away from the tree. In the spring there is a profusion of upright growth, called a candle.

12. The **bald cypress**, *Taxodium distichum*, is a native of East Texas. The bald cypress is a conifer that produces spherical cones. Mature trees reach a height of more than 100 feet. Native plants prefer to grow along the edges of streams and swamps. There they produce aerial roots called knees. The function of knees is not known.

The leaves are feathery and delicate. They grow on either side of the slender twigs. The leaves are also flexible. The leaves and the twigs are shed in the fall revealing a reddish brown-bark. Bald cypress is the only deciduous conifer.

13. The **green ash**, *Fraxinus pennsylvanica*, is a tall upright tree. The bark is ashy gray. The leaves have an opposite arrangement. The leaves are pinnately compound with seven to nine serrate leaflets. Leaflets are lance-shaped to elliptical in shape. The leaf is six to nine inches long. The upper leaf surface is dark green while the underside is silky-pubescent below. The petiole is reddish and smooth. The fruit is a one-winged samara that matures in September to October.

DATA TABLE 25.1 Characteristics of South Campus Plants

Scientific Name	Common Name	Pant Type (herbaceous, woody, shrub, tree)	Type of Lead Margin (smooth, lobed, serrated)	Type of Leaf Venation (parallel, pinnate, or palmate)	Leaf Arrangement (alternate/opposite)	Leaf Type (simple/compound)
1						
2						
3						
4						
5						
6						
7						
8						
9						
10						
11						
12						
13						
14						
15						
16						
17						

14. **Red yucca**, *Hesperaloe parviflora*, is not a yucca plant. Yucca plants have spikes on their leaves; these do not. It is a native of Central and Western Texas. It has a dark green rosette of long, thin leaves arising from the base. The leaves have fibrous threads at the edges. The venation is parallel. The plant is an evergreen and is a perennial. It produces a long spike of pink to red to coral bell-shaped flowers. The flowers attract hummingbirds.

15. **Desert willow**, *Chilopsis linearis*, is not a willow tree. The long narrow leaves resemble those of willows. The leaf margin is entire. The leaves are alternate in arrangement. Venation is pinnate. It is a delicate, small, deciduous tree native to West Texas and the Edwards Plateau. Its flowers are trumpet-shaped and sweetly fragrant. Colors range from light pink to light violet. Flowers are arranged in clusters. Flowers appear in summer to fall. Fruit contains winged seeds.

16. The **maidenhair tree**, *Ginkgo biloba*, is the only living member of the phylum Ginkgophyta. It matures as a large shade tree that is 80 feet tall by 60 feet wide. It has fan-shaped leaves, which are shed in the fall. There are male and female plants. The female plants produce a fruit that has a fleshy cover that produces a foul odor. This tree was planted in memory of Clyde Bottrell, professor of Biology. Some herbalists think that ginkgo bark can improve your memory. It has been proven that no positive effect has been found.

INDEX